Education, Community Engagement and Sustainable Development

Nicole Blum

Education, Community Engagement and Sustainable Development

Negotiating Environmental Knowledge in Monteverde, Costa Rica

Nicole Blum
Development Education Research Centre
Institute of Education
University of London
36 Gordon Square
London WC1H 0PD
United Kingdom

ISBN 978-94-007-2526-3   e-ISBN 978-94-007-2527-0
DOI 10.1007/978-94-007-2527-0
Springer Dordrecht Heidelberg London New York

Library of Congress Control Number: 2011942519

© Springer Science+Business Media B.V. 2012
No part of this work may be reproduced, stored in a retrieval system, or transmitted in any form or by any means, electronic, mechanical, photocopying, microfilming, recording or otherwise, without written permission from the Publisher, with the exception of any material supplied specifically for the purpose of being entered and executed on a computer system, for exclusive use by the purchaser of the work.

Printed on acid-free paper

Springer is part of Springer Science+Business Media (www.springer.com)

Preface

This book was originally written as a doctoral thesis in the Department of Anthropology, University of Sussex, UK. As such, it needed to conform to the requirements and standards of the degree, the university and the wider discipline. Since, like most doctoral work, it was unlikely to be read in that form by anyone other than my supervisor, examiners, and a few (very kind) family, friends and colleagues, I have thoroughly revised the text in an attempt to make it appealing and accessible to a much wider audience. Hopefully, both academics and practitioners in environmental education and related fields will now find the discussion and arguments useful in their thinking and practice around education, development, environment and sustainability.

There are a great many people to whom I am deeply indebted for their help and support in both completing my PhD research and finishing this book. Although it is impossible to give thanks to everyone by name, I hope that many friends and colleagues will be pleased to see their influence in the final product.

Firstly, for its financial support for this research between 2002–2004 while I was a doctoral student, my thanks to Universities UK and the Overseas Research Student Scheme. For professional support, supervision and guidance, I owe thanks a great many people at both the University of Sussex and the Institute of Education, University of London, where I am currently based. Chief among these are colleagues in the Development Education Research Centre, and especially Douglas Bourn and Clare Bentall, who have provided thought-provoking discussion as well as an encouraging work environment in which this book could finally be finished. My thanks also to Tim Wallace at North Carolina State University (USA), who first introduced me to Costa Rica and to doing fieldwork in 2001.

In Costa Rica, I am grateful to the many people who participated in or assisted with my fieldwork, and who have read and commented on the work in later stages.[1]

---

[1] Throughout the book, I have used the real names of research collaborators only when their views or comments were already published, and therefore in the public domain. All other names have been changed to protect privacy. Material taken from these interactions is denoted in one of

In the Central Valley, this includes contacts and collaborators at the Ministry of Education (*Ministerio de Educación Pública*; MEP), the Ministry of Environment and Energy (*Ministerio de Ambiente y Energia*; MINAE), the *Instituto Costariccense de Turismo* (ICT; Costa Rican Tourism Institute), the *Universidad de Costa Rica, Universidad Nacional,* and *Universidad Estatal a Distancia,* the *Instituto Nacional de Biodiversidad* (INBio; National Biodiversity Institute), the Tropical Science Center, the Organisation for Tropical Studies, WWF-Central America, and the Costa Rican law group CEDARENA. In the Monteverde region, my thanks goes to the many individuals and organisations who so kindly welcomed me into their classrooms and meeting spaces, participated in the research, answered my numerous questions with good humour, and more generally shared their time, energy and insights with me.

I am particularly indebted to the educators, administrators and other staff at the Monteverde Reserve (*Reserva Biológica Bosque Nuboso Monteverde*), the Santa Elena Reserve (*Reserva Bosque Nuboso Santa Elena*), the Monteverde Conservation League, the Monteverde Institute, the state primary schools in Santa Elena and Cerro Plano, the *Colegio Técnico Profesional de Santa Elena* (state secondary school) in Santa Elena, and the Cloudforest School. Their collaboration and assistance, both throughout the fieldwork year and since, has been absolutely invaluable. My thanks also to the Monteverde Institute whose director at the time kindly gave me permission to cite data from their 2002 survey of the community, which proved helpful to constructing an overview of the local context. Also in Monteverde, but on a more personal level, I am deeply indebted to Virginia Kennard, Cindy Olivares Rodriguez, Dennis Gomez, and Patricia Jiménez for their friendship and support both during the fieldwork year and since.

In the process of revising the text for this book, I have also benefitted from feedback from anonymous reviewers for the journals *Environmental Education Research*, the *International Journal of Educational Development*, and *Ethnography and Education* – each of whom helped me to revise, re-think, and strengthen pieces of this research for publication.

Finally, I would never have been able to do the research and writing at all without the support of family and friends. My deepest appreciation and thanks to my parents, Udo and Mary Ann Blum, and my sister, Amy Blum Grady, who have stood by me through thick and thin. The same goes to the many friends and fellow doctoral students who provided support and sympathy over many, many cups of coffee during the last several years. Special thanks also to Nicky Swetnam and Alice Taylor who made the long journey to visit me in Monteverde in May of 2003 and also allowed me to share some photos from that trip in this book. Finally, my most heartfelt thanks to James Facey, my husband and dearest friend, to whom I am forever grateful for his steadfast support.

---

two styles:direct quotations taken from recorded interview transcripts are placed in inverted commas; paraphrased passages taken from my interview notes are in italics. Although passages in italics do not represent direct quotations, they are true to the intent of the conversation in which they took place. All translations of interview materials and Spanish-language publications are mine.

Maps by Lucid Design, Brighton, UK

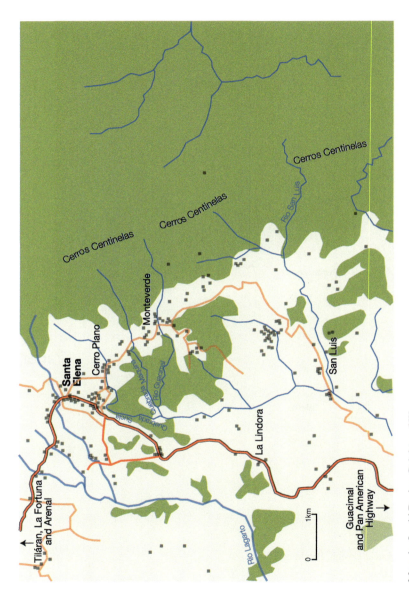

Contents

1	**Introduction**	1
	Environmental Education in Latin America	2
	Nature of the Research	5
	Central Arguments	7
	The Need to Widen Perspectives	7
	The Need to Understand Environmental Learning as a Process and Learners as Active Participants	8
	The Need to Explore Learning in Context	8
	International Policy on Environmental Education	9
	Education, Learning and Social Change	11
	Exploring Perspectives on Environmental Learning	13
	Structure of the Book	16
	References	18
2	**Education, Environment, Development and the Costa Rican State**	23
	Historical Development of National Education Discourses and the State Education System	27
	Organisation of the National Education System	29
	Implementation and Effectiveness	31
	State Environmental Management, the Ecotourism Industry and Scientific Research	33
	State-Funded Environmental Education	38
	Schools	38
	National Conservation Areas	40
	Defining and Implementing Environmental Education in the State System	41
	Regional and International Influences	44
	References	48

3 Environmental Education in Schools ... 53
Welcome to Monteverde ... 54
Formal Education in Monteverde ... 58
Environmental Education in Monteverde's Schools ... 59
State Primary Schools and *Temas Transversales* ... 60
   Training for Environmental Education ... 62
State Secondary Schooling ... 64
   Environmental Learning in the Ecotourism Programme ... 66
   Teachers' Perspectives on the Curriculum ... 67
The Cloudforest School ... 69
   Defining and Implementing Environmental Education ... 71
   Relationships to the State and the Community ... 75
Teaching Science or Cultivating Values? ... 77
References ... 78

4 Environmental Education and Conservation Organisations ... 81
Early Settlement and the Growth of Local Conservation ... 83
Science, Conservation and Local Environmental Management ... 87
NGOs and the Growth of Local Environmental Education ... 90
A Tale of Two Reserves ... 92
   Promoting Ecological Knowledge ... 94
   Educational Approach and Impacts ... 97
   Training the Next Generation of Educators ... 100
   Educational Approach and Impacts ... 103
Diverse Approaches, Diverse Positions ... 104
Environmental Knowledge and Community Relationships ... 106
References ... 106

5 Environmental Knowledge in Public Spaces ... 109
Local Development Issues ... 112
Community Organisation and the Importance
of Public Spaces for Education ... 115
Local Styles of Debate and Negotiation ... 116
   Discourses of Sustainable Development… ... 117
   … and Competing Discourses of Development ... 122
Participation in Local Organisations ... 124
Community Development: Ecotourism
vs Environmental Protection? ... 127
Competition and Cooperation in the Local Ecotourism Industry ... 131
Negotiating Environmental Knowledge in the Community ... 135
Conclusions ... 138
References ... 139

6 Conclusions ... 141
Contributions to Existing Research ... 145
Environmental Learning in Latin America ... 147
Moving Forward: Areas for Future Research
and Policy in Environmental Education ... 149
References ... 153

Appendix: Images of Monteverde ... 157

Index ... 161

About the Author

Nicole Blum is a Lecturer in the Development Education Research Centre at the Institute of Education, University of London (UK). Since completing her PhD (Anthropology) at the University of Sussex in 2006, she has been actively engaged in research on environmental education and education for sustainable development, and educational responses to climate change, as well as internationalization and global learning in higher education. Her work also includes an interest in the ethnography of education, the anthropology of development, access to and participation in education, and Education for All. In addition to teaching on an online MA programme in Development Education, she is also currently coordinating a research project which aims to develop and evaluate methods to embed learning about global and development issues within undergraduate courses in pharmacy, veterinary science, and medicine. Over the course of her career she has worked and conducted research in Costa Rica, Guatemala, India, the United States and the United Kingdom.

Chapter 1
Introduction

Abstract Environmental education has been at the centre of international and national policies of sustainable development for the last several decades, and has stimulated significant debate regarding both its inclusion in educational programming and proposed methods for implementation. Research has given critical attention to diverse theories and practices of environmental education, but has tended to take a narrow focus on specific curricula and policies or on activities within strictly defined sites such as classrooms or natural areas. The field has also largely been focused on discussions and initiatives in industrialised countries, and neglected to explore perspectives, policy and practice elsewhere in the world. In contrast, this chapter introduces a research study based on anthropological fieldwork that explored environmental education and learning in the community of Monteverde, Costa Rica. In particular, the research set out to both explore initiatives taking place in formal education, in programmes organised by non-governmental organisations, and through engagement in public education spaces, as well as the relationships between them and the wider community.

Keywords Costa rica • Environmental education • Ethnography • Learning • Sustainable development • Community • Forest conservation

At first glance, environmental education may appear to be a rather simple idea – basically, increasing public awareness of environmental issues – but the term is commonly used to denote a wide variety of ideas and practices. These diverse understandings have catalysed significant debate from practitioners, researchers and policy-makers on a number of key issues. At one end of the spectrum, for instance, are perspectives that claim that environmental education should centre on teaching about environmental issues through the natural sciences, while at the other are those that argue for a more holistic approach that combines learning about environmental concerns with discussion of social issues, economics and development. Practices of environmental education are also varied: it can be found in classrooms, public meetings and conference halls, as well as in forests, on beaches, and in museums and botanical gardens. Learners, too, are diverse; they may be young or old, privileged students or impoverished agricultural workers. Despite this diversity of ideas and practices, however, environmental education is either implicitly or explicitly underscored by an ethical imperative to protect and preserve the natural world.

International attention to these issues has been highlighted by the current UN Decade for Education for Sustainable Development (2005–2014), as well as increasing concern about climate change and its impacts around the world. The Decade follows in the footsteps of international agreements such as *Agenda 21*, which famously called for a 're-orientation' of all education towards sustainability (UNCED 1992, chap. 36). Many international organisations, national governments, and nongovernmental organisations (NGOs) have devoted significant resources to raising awareness of these issues, as well as to developing new curricula and supportive teaching resources to address them.

Over the last several decades, a growing body of research has also sought to assess the practices and potential of environmental education in a variety of geographical locations and social contexts. Influential research communities in the UK, Europe and North America have historically provided the bulk of analytical and case study material within the existing literature. However, new research in a number of other nations has begun to provide much greater breadth to the literature's geographical focus (cf. Pellegrini Blanco 2002; Lai 1998; Kwan and Lidstone 1998; Ouyang 2000). Undoubtedly, this addition of more diverse perspectives is important to developing a much deeper understanding of the variety of ways in which environmental learning is conceptualised, practiced and critiqued around the world. It also provides important opportunities for questioning the often dominant 'Western' or 'Northern' ways of understanding education, learning and the relationships between human beings and the natural world. As Australian researcher Annette Gough noted more than a decade ago: 'This silencing of non-Western perspectives in the dominant discourses of environmental education is an ongoing issue and a challenge for the future of environmental education' (Gough 1997: 36).

Environmental Education in Latin America

Latin American traditions of employing education as a tool for social change – and especially through the region's long history of liberation theology and popular education initiatives – as well as a complex history of relationships between Latin American states and other nations (particularly the United States), make it especially important to attend to perspectives on environmental education and sustainable development in the region. This book is therefore an attempt to add to the existing literature on these topics in Latin America (cf. Arnove et al. 1999; Freire 1972; Honey 1994; Morales-Gómez and Torres 1992; Stromquist 1992), as well more as generally to international understandings of environmental education and sustainable development.

In doing this, it is important to recognise that the development of environmental education thinking and practice in Latin America has followed a somewhat different trajectory to that which has taken place in the UK, Europe and North America. As other authors have noted, the understandings of environmental education expressed in international statements in the 1970s often had little resonance with educators or policy makers in either Latin American nations or other 'developing' countries.[1] Mexican environmental education researcher Edgar González Gaudiano argues, for instance, that the Stockholm Declaration (1972) promoted 'a type of education which fit with understandings of the environmental concerns faced by the industrialised world; in other words, an understanding that these problems were solely ecological in nature' (1999: 13). Such conceptualisations neglected to account for either the differing nature of environmental concerns facing Latin American nations or the larger economic and political concerns – including military or civil conflict – which they faced in trying to address them. In contrast, 'the environmental situation in Latin America is not the result of abundance and waste, but rather the lack of basic necessities, which is in turn the cause of malnutrition, illiteracy, unemployment and ill health' (González Gaudiano 1999: 14).

Similarly, while early international discussions of environmental education led by industrialised nations in the 1960s and 1970s emphasised the importance of learning about the natural sciences and promoted largely transmissive approaches to teaching about environmental concerns, discussions within Latin America instead often drew strong linkages between environmental education and wider social and political movements and concerns. At a regional workshop on Environmental Education within Secondary Teaching in Chosica, Peru in 1976, for instance, the attendees – educators and students from Cuba, Panama, Peru, Venezuela, Argentina and Brazil – developed a definition which viewed environmental education as:

> '…. a continuous activity by which the educational community takes into account the global reality, the types of relationships that human beings establish between themselves and nature, the problems which result from those relationships, and their fundamental causes. It is developed through practice which links the learner to the community, values and attitudes that promote behaviours directed towards the transformation of that reality, in both environmental and social terms, and develops in the learner the skills and aptitudes needed for that transformation'. (Teitelbaum 1978: 51 cited in González Gaudiano 1999: 14)

The nature and potential role of environmental education in social change have continued to characterise discussions in the region more recently, as represented by the many regional workshops which have taken place alongside – and sometimes in response to – international meetings. Topics of discussion at the Latin American Congress on Environmental Education in Guadalajara, Mexico in 1992 (just 6 months after the Rio Summit), for example, included projects to explore and

---

[1] Throughout the book, I use the terms 'developing' and 'developed' with caution because such categories are highly problematic, and often inaccurate (cf. Sachs 1999; Rahnema and Bawtree 1997; Grillo and Stirrat 1997). Nevertheless, both terms are commonly used in contemporary discussions of global relationships and provide a useful way of highlighting the inequalities that are often a feature of them.

address social inequality, poverty, communication and access to media, the development of environmental education legislation, gender issues, and the cultures of indigenous populations (González Gaudiano 1999: 22).

These themes resonate strongly in Costa Rica, where this research was conducted, as do many of the challenges faced by environmental educators elsewhere in the region. The case of Costa Rica is significant to the broader study of environmental education, community engagement and sustainable development both in Latin America and around the world because of the importance of both education and environmental protection to the nation's economy and identity. Innovative efforts in environmental protection for the last several decades have made the country one of the acknowledged world leaders in efforts to achieve sustainable development and environmental management. In the mid-1990s the nation was even offered up to the international community as the 'ideal international test case for sustainable development projects' (Figueres Olsen 1996). Its reputation as 'the green republic' has brought with it both international attention and substantial economic benefits. In addition to profits from a successful ecotourism industry, international organisations have also invested heavily in conservation projects, research, and innovative environmental management schemes, as well as taking part in national policy formation and campaigning. The scale of this international attention is quite impressive: a 1995 World Resources Institute study, for example, concluded that there were more conservation projects in tiny Costa Rica than in all of Brazil (cited in Boza et al. 1995: 684).

Perhaps unsurprisingly, environmental education is a significant topic of debate and discussion within Costa Rica, and is explored within a substantial literature that outlines key concepts, policy and practice. This literature is largely unknown outside the country, however, and although a few international researchers have given attention to environmental education in the past, this has usually taken the form of general descriptions of programmes linked to broader national conservation and sustainable development efforts (cf. Evans 1999; Honey 1999). As a result, an in-depth examination of the links between environmental learning and sustainable development in Costa Rica has been missing from international discussion and analysis.

This book therefore set out to draw together a number of key strands of theory, concepts and understandings from international research on environmental education and learning, and to connect these to the everyday realities of a community of educators, students, parents, and policy makers in Costa Rica. In order to do this, I draw on a range of related research and writing conducted under various terms and definitions, including not only environmental education, but also education for sustainable development, development education, global education, and science education, among others (cf. Bourn 2008a; Huckle and Sterling 1996; Pike 2008). However, while recognising the importance of these different concepts and the debates which surround them, I use the term 'environmental education' (in Spanish, *educación ambiental*) throughout the book because it is the term most commonly found in Costa Rica, where it is used to describe a wide range of types of educational programming.

Nature of the Research

More specifically, this book represents the personal journey I took during the year I spent living and working in Monteverde (September 2002–September 2003), a rural mountain community in Costa Rica. As a doctoral student in anthropology, my work used a broadly ethnographic approach, which followed the discipline's tradition of conducting extended fieldwork and using key methods such as participant observation and interviewing (cf. Bernard 1988; Hammersley and Atkinson 1995). In addition to formally interviewing a number of local residents – including environmental educators, teachers and administrators in private and state schools, development project co-ordinators, directors and staff of NGOs, conservationists and protected area managers, scientists and other researchers, tourism business owners and government officials – who kindly agreed to speak to me, I was also invited to take part in a range of community events and activities which gave me a sense of the life of the community. Perhaps the most significant of these exchanges took place through arrangements with school teachers and with environmental educators based within conservation organisations. Through these, I was able to provide some support for environmental education programming in exchange for opportunities to observe projects and informally interact with both educators and students. In some cases, this assistance took the form of lesson planning or preparation of materials, in others it involved overseeing student activities or contributing to group discussions. This exchange of my time and energy for research access was also an important part of my commitment to try to 'give something back' to the community – no matter how small a contribution it usually was.

My intensive work in the Monteverde region was also complemented by interviews and visits with policy-makers, educators and conservationists in the capital city, San José, and in other areas of the country. These proved important to understanding the role of environmental education in Costa Rica, provided useful comparisons to Monteverde, and also helped me to understand the networks and interconnections (and in some cases, disconnections) between the numerous individuals and organisations involved in promoting environmental learning in the country.

In terms of style and approach, therefore, the following chapters rely heavily on what might be called 'ethnographic storytelling'; in other words, using narratives and experiences which research participants shared in order to highlight key points. This is because I recognise the importance of allowing those individuals – all of whom are far more knowledgeable about their subject than me – to give voice to their own ideas and perspectives. However, in an effort to protect their privacy, throughout the book I have used the real names of research collaborators only when their views or comments were already published, and therefore in the public domain. All other names have been changed.

Although much of the discussion and analysis which follows is necessarily rooted in my own thinking and experiences, I also made a concerted effort to involve members of the community in the process of analysing and writing up the research, initially as a doctoral thesis at the University of Sussex, UK (Blum 2006). Before leaving the community in September 2003, for example, I gave a public presentation to ask for feedback on the research process and on some of my preliminary conclusions. The ideas and comments community members shared with me on that occasion were highly significant to the final shape of the analysis. This discussion has continued since my departure, although on a much more limited scale, with some of the key research participants kindly taking the time to give their comments and suggestions on draft analyses and papers.

It is also worth noting here that while conducting the research I was keenly aware of my own identity as a researcher and of its potential impacts. Monteverde has received significant attention from researchers (mostly, but not only, natural scientists) since the 1970s. Unfortunately, interactions with previous researchers have not always been positive, and some community members were understandably cautious about taking part in this research. My own nationality (US American), gender (female), ethnicity (white European), professional status (at the time, a doctoral student in anthropology), and language abilities (English and Spanish) undoubtedly had both positive and negative influences on my relationships with particular organisations and individuals. Overall, however, I found myself welcomed into a community of active individuals and organisations engaged in intensive debate over the present and future well-being of their community. It is my sincere hope that the following chapters provide a coherent account of both the many examples of innovative thinking and practice that were shared with me, as well as of the key challenges the community continues to face.

Given the length of time which has passed since the fieldwork phase of the process, however, it is also only right to emphasise that the descriptions and stories included in this book provide an account of both particular organisations and individuals, as well as the community as a whole, during a specific moment in time. More recent changes in the global economic climate, among other things, have had significant impacts in Monteverde just as they have across the rest of the world, and I have tried to provide updates where appropriate. Nevertheless, despite such changes since 2003, I believe that the key issues raised by this research continue to resonate both in Monteverde and in other communities elsewhere.

This is because, while the specific individuals and organisations involved in education, conservation and development in the community may have changed over time, the research identified three key tensions which feature in both historical and contemporary perspectives and practices surrounding local environmental education. These included debates about (i) what constitutes the most 'appropriate' content of programmes (and particularly whether these should emphasise teaching of the natural sciences or engagement with social issues), (ii) how programmes should be organised (e.g. in formal curricula, in nonformal education, via community NGOs or businesses), and (iii) who has the power to make those decisions. Certainly, these themes continue to be a subject of intense discussion within Latin America more broadly, and are also likely to remain relevant in other 'developing' country contexts.

On a more practical – and perhaps even more urgent – level, issues related to conservation, environmental management and sustainable development continue to be highly significant both in Monteverde and in many other communities around the world. Given its geographical location and altitude, the Monteverde region has also already begun to experience the effects of climate change, including the loss of several rare endemic species. Exploring these topics in a community that is in many ways on the 'front line' of environmental management and sustainable development therefore seems important, both in order to understand what has happened in the community in the past, as well as to consider what lessons this might have for others facing similar challenges in the present day.

Having said this, I do not mean to suggest that Monteverde is a model of good practice which other communities should simply try to replicate (although I sincerely hope that readers will find useful ideas to inform their own work or perspectives), because the tensions surrounding education, environmental management and sustainable development are likely to play out quite differently in diverse contexts. However, the importance of ecotourism, conservation and research in the local economy gives environmental education a particularly central role in debate and discussion in Monteverde. This in turn makes it a useful setting for developing an understanding of the impacts of community dynamics on perspectives and practices of environmental education and sustainable community development.

Central Arguments

In very broad terms, the book as a whole therefore sets out to make three interrelated arguments:

The Need to Widen Perspectives

There is ample evidence from research and practice to suggest that educational programming can have an impact on attitudes, behaviours and skills related to environmental management and sustainable development (cf. Lozt-Sisitka 2004; Dillon et al. 1999; Palmer et al. 1999; Jaritz 1996; Kwan and Lidstone 1998). However, research in the area has tended to be somewhat limited in scope. Firstly, it has tended to focus on single sites, such as classrooms, NGO programmes, or strategies in nature areas. While this work is certainly important, taking such a narrow focus has meant that the relationships *between* theory and practice or *between* different educational sites have often been left unexamined. It has also meant that the ways in which practices and perspectives of environmental education are deeply embedded in particular social, historical and economic contexts have not often been fully explored.

Secondly, there has been a tendency for research to rely on short-term data collection techniques (e.g. surveys and questionnaires; short-term fieldwork) to investigate complicated issues such as attitude formation and behavioural change. While these methods can help to shed some light on environmental knowledge and understanding, they are generally able to say little about how or why knowledge and attitudes are applied (or not) in the everyday lives of the people involved. They also cannot fully account for the diversity of responses to learning about environmental topics, and how these might relate to sustainable community development in practice.

The Need to Understand Environmental Learning as a Process and Learners as Active Participants

Research, policy and practices of environmental education have in the past often tended to treat 'learning' as a simple process of knowledge transfer, rather than as a part of wider processes of social interaction and exchange. To put it more simply, it has often been assumed that new inputs (e.g. curricula, textbooks, training programmes) will result in significant changes to either an individual or a system. However, many decades of educational research have illustrated how complex learning processes can be, and have also roundly critiqued so-called 'transmission' theories which suggest that learning is the simple transfer of information from teacher to student (cf. Illeris 2007). Instead, learning is better understood as a complex and continuous process of acquisition, accommodation, interpretation and capacity change – all of which are in turn influenced by a number of individual and social factors, including relationships to particular contexts and communities.

As other authors in the field of environmental education have recently suggested, much more work is therefore needed to understand these complex learning processes more fully (cf. Rickinson 2001; Scott and Gough 2003; Heimlich and Ardoin 2008; Rickinson et al. 2009). A growing number of researchers have also begun to argue that learning which encourages skills such as critical thinking and problem solving, as well as encouraging flexibility and adaptability, are important to addressing the challenges of sustainable development and climate change, and of living in a rapidly changing world more generally (cf. Bourn 2008b; Gough and Scott 2007; Bangay and Blum 2010). These arguments open up important conversations about the multiple and diverse approaches to environmental education which are represented in research, policy and practice, as well as their potential impacts on individual learners and communities.

The Need to Explore Learning in Context

Finally, as with every kind of educational programming, practices and perspectives of environmental education are deeply embedded in particular social, historical and economic contexts. Ideas about what constitutes 'appropriate' programme content or teaching methods, for example, are closely related to the political and social positions of the individuals and organisations involved in its promotion. What kinds of knowledge does (or should) environmental education include? Why are some kinds of knowledge prioritised while others are marginalised? What kinds of approaches to learning are (or should be) promoted? What are the relationships between environmental education and action in support of sustainable development? Diverse actors in diverse contexts are likely to answer these questions in very different ways, and their perspectives will have an impact on opportunities for collaboration as well as the potential for conflict within and among communities seeking to achieve sustainable development.

It is these gaps in existing research and writing that I try to address in this book by giving attention to diverse educational settings – including schools, community organisations, and public educational spaces – in one community in Costa Rica, as well as to the interactions *between* them, and to the complex social and environmental contexts in which they are located. I have not, it should be noted, set out to illustrate how to *do* 'good environmental education' because numerous references and guides already exist which attempt to do this.[2] Rather, this book sets out to explore the ways in which knowledge about environmental issues moves through a community, and how this is related to local efforts to promote sound environmental management and sustainable development. In doing so, I hope that it provides a useful case study for other communities and researchers interested in, or working towards, similar goals.

International Policy on Environmental Education

Before moving to the specific case of this research, it is worth pausing to outline some of the key international policies and discussions related to environmental education. In order to do this, I will draw largely on literature that has been published in English, including literature from the UK, North America and Australia as well as international organisations such as UNESCO and UNEP (United Nations Environment Programme). This reliance on English language literature should not be read as a suggestion that understandings of environmental education and learning from 'Western' perspectives are the most useful or important. Instead, it is an indication both of my own positioning as a UK academic, as well as of the historical dominance of these perspectives within environmental education and related fields. Nevertheless, I hope that this body of work – limited as it may be in some respects – provides a helpful way to stimulate wider international discussions around shared debates and concerns.

In general terms, the international environmental education movement is commonly supposed to have arisen in Europe in the 1960s as a result of growing public

---

[2] See, for example, numerous resources on the websites of the UK's Council for Environmental Education and the North American Association for Environmental Education.

concern over the state of the natural world, as well as the growing interest of the international scientific community in ecosystem and species preservation (cf. Palmer 1998; Smyth 1995). With its roots in the 'Western' scientific community, early environmental education relied heavily on a style of public education and awareness-building that emphasised learning in the natural sciences, including biology, botany and ecology. As a result, these educational programmes tended to emphasise the interests, concerns and perspectives of industrialised countries. Topics of interest included, for instance, the scientific study of food and agriculture, tropical forests, biological diversity, desertification and drought, water management, energy, climate, solid waste and sewage management, and population growth (Palmer 1998).

Support for education about such environmental concerns within many international organisations grew steadily in the following decade, as evidenced by statements from the United Nations Conference on the Human Environment in Stockholm, Sweden (1972), the establishment of the International Environmental Education Programme (1975), and the publication of The Belgrade Charter (1975). As with previous statements, the language of the Belgrade Charter strongly emphasised the need for greater knowledge of 'the environment and its associated problems' (UNESCO-UNEP 1976), and focused on raising awareness of the environmental damage caused by human activity. A follow-up conference, the First Intergovernmental Conference on Environmental Education, was hosted by UNESCO in Tbilisi, Georgia, USSR in 1977. The final report of the conference – the Tbilisi Declaration – contains recommendations for implementation of environmental education in formal and non-formal education, as well as a framework for international cooperation that is still in use today (UNESCO 1977).

By the 1980s, however, discussions in the field had begun to change in order to account for both the emerging idea of 'sustainable development' as well as demands for attention to more diverse voices and perspectives coming from 'developing' countries (sometimes also labelled as the Third World or global South). The World Conservation Strategy (IUCN 1980), for example, utilised the term 'sustainable development' for the first time and strongly emphasised the links between conservation and (economic) development in 'developing' countries. The growing emphasis on 'sustainability' continued with the United Nations Conference on Environment and Development (commonly known as the 'Earth Summit'), held in Rio de Janeiro in 1992. Central to the publications coming out of the conference were both The Rio Declaration and Agenda 21, a key set of plans and international agreements aimed at achieving global sustainable development in the twenty-first century. Education was given a central role in the plans outlined by Agenda 21:

> 'Education, including formal education, public awareness and training should be recognized as a process by which human beings and societies can reach their fullest potential. Education is critical for promoting sustainable development and improving the capacity of the people to address environment and development issues. While basic education provides the underpinning for any environmental and development education, the latter needs to be incorporated as an essential part of learning.' (UNCED 1992: sect. 36.3)

Some educational researchers have since argued that the role for education outlined in Agenda 21 represents a significant change to earlier understandings of environmental education. This is due to its much broader attention to basic education and social concerns such as human rights and gender inequality:

> 'The overall intent had moved from environmental protection and pollution reduction to addressing the needs of both environment and society. The goal shifted to finding a realistic and balanced approach to environmental protection while alleviating human suffering and the ravages that accompany poverty.... While it is clear that environmental education was never devoid of social and economic concerns, there is nevertheless a clear shift of emphasis implicit in the notion of sustainable development.' (McKeown and Hopkins 2003: 119–120)

A new vocabulary surrounding education about environmental concerns also emerged in response to these debates. While the term 'environmental education' dominated policy language and practitioner vocabulary for several decades – and is still the most meaningful term for many researchers and educators around the world today – by the late 1980s some educationalists had also begun to propose new terms to describe their work. Advocates of these alternate approaches – including 'education for sustainable development', 'socially-critically environmental education' and 'grass-roots environmental education', among many others – argued that early proposals of environmental education had been too limited in scope both because they were rooted in particular 'Western' perspectives and because they focused 'too narrowly on the protection of natural environments (for their ecological, economic or aesthetic values), without taking into account the needs and rights of human populations associated with these same environments' (Sauvé 1996: 7). They argued instead for the inclusion of related issues such as peace, human rights, gender inequality, and cultural identity as part of the process of achieving sustainability. Debates about the relationships between education, environment and development since that time have given significant attention to the relative merits and limitations of these various approaches to education and their potential to catalyse individual and social change. Throughout this book I will argue that although the dialogues surrounding these terms and definitions have come from a variety of angles and represent a range of interests, at their core they largely turn around fundamental questions about the role of education in social change, especially as this relates to environmental management and sustainability.

Education, Learning and Social Change

In particular, within environmental education research, policy and practice there continues to be a clear tension between perspectives that emphasise teaching of science concepts and those that seek to more actively link environmental and social issues. This tension mirrors the historical development of international discussion and policy in the field (outlined in the previous section) and in Latin America particularly (see González Gaudiano 2007). It is also often evident in everyday practice, with some educators and organisations advocating a focus on science learning, others arguing for the integration of efforts in environmental management and human development, and still others working to find a balance between the two.

In simplified terms, advocates of more science-oriented styles of environmental education (educators, organisations, policy makers, parents and others) tend to argue for an emphasis on teaching about ecological and biological issues, and claim that when individuals are taught about these issues they will learn to love – and therefore be inspired to protect – the natural world from destruction. This approach is supported by work in areas such as 'environmental interpretation' (cf. Ham 1992) and is often a feature of educational programmes operating in parks and other protected areas.

At the other end of the spectrum, supporters of programmes with a strong social-values orientation, on the other hand, claim that environmental issues cannot be studied in isolation, but should instead be taught in relation to human needs and activities. This perspective encourages the development of critical thinking about issues such as human rights, peace, poverty and inequality, and writings tend to focus more explicitly on values and responsibilities. Researcher Arjen Wals has argued, for example, that environmental education should be 'a learning process that seeks to enable participants to construct, transform, critique and emancipate their world in an existential way' (1996: 301).

One fundamental question for the field is therefore: Is the central role of educational programmes to teach environmental 'facts' and encourage particular 'environmentally responsible' behaviours or is it to encourage participants to develop skills of analysis and critique which they can use to understand and respond to the world around them?

In educational terms, the first perspective tends to view learning as a process of 'transmission' – i.e. of facts and ideas to a receptive audience – which has planned/directed outcomes. These might include inspiring learners to participate in specific activities such as forest conservation or recycling, or to take part in advocacy work or campaigns. The second perspective, on the other hand, tends to view learning as a process of transformation/personal development in which the outcomes are often uncertain. The aim of these educational programmes is not to promote specific, pre-determined actions or behaviours, but rather to encourage learners to develop the skills to understand the world around them and to make decisions about their actions and behaviours based on their individual beliefs, values and needs.

In practice, of course, the distinctions between these perspectives are often blurred, and environmental educators may employ a whole range of teaching approaches, perspectives and activities in their practice. Recent work suggests that there are also researchers with views that could be placed along the entire spectrum between these two orientations (cf. Courtenay-Hall and Rogers 2002; Kollmuss and Agyeman 2002).

Given this diversity of perspectives and practices, therefore, it is difficult – and perhaps not even desirable – to make an assessment of their relative value. The results of such an assessment are also likely to vary widely depending on the specific context being discussed, from whose point-of-view it is conducted, the learners involved, and the relative influence of those involved in the process. For those reasons, this research instead set out to explore and better understand the dynamic interactions between diverse educators, organisations and members of one particular community, especially as they relate to debates on learning, change and (sustainable) community development.

Exploring Perspectives on Environmental Learning

Examining these large and multidimensional issues related to education, learning, the environment and sustainable development requires a correspondingly broad approach to research. The field of interest must include a number of diverse individuals and organisations, the inter-relationships between them, the interactions between different kinds of educational experiences (formal, informal, etc.), and the dynamic social and economic relationships in which those diverse perspectives and activities are embedded. This kind of approach draws on work from a number of fields, including not only environmental education, but also mainstream educational research, as well as perspectives and methodological approaches from anthropology.

Anthropologists have always taken an active interest in education and its links to wider social life, especially during moments of social change. This was as true during the era of colonial encounter as it is now in the context of rapid globalisation. Since the late nineteenth century, anthropological research has also contributed strongly to understandings of pedagogy, the school curriculum and childhood (Eddy 1985: 84). Research in anthropology, as well as education and sociology, raises key questions about the ways in which education is bound up in particular social and cultural contexts (cf. Spindler 1955; Wax et al. 1971; Williams 1961; Willis 1966). These questions continue to be relevant in the present day.

From its beginnings, anthropological research has given attention to both 'education' (commonly defined as formal efforts, such as schooling, within the group under study) and 'cultural transmission' (usually understood as informal efforts, often within the community or the household) within the exploration of diverse cultures and societies. Since its formal recognition in the US in 1954, the sub-discipline of the 'anthropology of education' has provided even more in-depth focus on topics as varied as school and curriculum reform (cf. Hess 1999), race, ethnicity and class in school settings (cf. Stocker 2005; Horvat and Antonio 1999; Cousins 1999), and indigenous language education (cf. Henze and Davis 1999), as well as the nature of diverse national education systems (cf. Singleton 1967; Ouyang 2000) and the social and cultural contexts of learning (cf. Street 2001; Lave and Wenger 1991).[3]

Approaches to informal learning in the anthropology of education have also begun to extend the borders of research to examine 'free spaces' in which 'deep, sustained community-based educative work, outside the borders of formal schooling' can be observed (Fine et al. 2000: 132). Borrowing from social psychology theory, this work explores the ways in which people use social spaces – such as geographic sites provided by community centres or sites connected to a shared religion, ethnicity or language – to contest public representations, to form or re-form identities, and to critique existing structures of power (cf. Boyte and Evans 1992; Fine and Weis

---

[3] Given this sensitivity to the role of both formal and informal education in social transformation, it is perhaps surprising to note that anthropological research has yet to devote sustained attention specifically to environmental education.

1998). This approach re-enforces the long-acknowledged idea that: 'Education does not take place just in schools, as anthropologists well know. It occurs at dinner time, in front of the television set, on street corners, in religious institutions, and in coffee shops' (Fine et al. 2000: 131). It also resonates with the strong tradition of social research on education in Britain, and especially with discussions of 'the hidden curriculum' and 'informal education', such as those popularised by Paul Willis (1966).

Environmental education researchers have, of course, also been attentive to both formal and informal sites for learning, perhaps partly because programmes are frequently implemented in 'informal' settings such as museums and national parks. As early as 1995, in the inaugural issue of the influential journal *Environmental Education Research*, environmental educator John Smyth commented:

> 'Formal education in schools and further and higher institutions, carried out by identifiable people trained for the purpose, is important, and often sets standards by which education is defined and judged. However, people also learn how to behave towards their environment in their homes and communities, during leisure activities, in the workplace, and from relatives, peer-groups, cultural influence, the mass media, advertising and the public example set by those in authority (as well as from legislative and fiscal measures). Different influences predominate at different times of life and in different circumstances: collectively they are a sustained and lifelong learning experience.' (Smyth 1995: 6)

Researchers publishing work in the same journal have more recently offered even more challenging discussion about sites of environmental learning outside of formal schooling. Rather than prescribing to the older distinctions of 'formal', 'informal' and 'non-formal' education, some researchers within the field have argued for an understanding of 'free-choice learning'. According to John Falk (2005), the distinctions between so-called formal, informal and non-formal education were first drawn by educators working in adult education and international development, and were used as a way of distinguishing between the differing educational experiences available in so-called 'developing' and 'developed' countries. The terms were then taken up by museum directors and environmental educators in the 1970s in order to mark the perceived differences between their programmes and those happening in school environments (Falk 2005: 271). However, as Falk and a number of other researchers have argued since the 1980s (cf. Kola-Olusanya 2005; Koran et al. 1983; Falk and Dierking 2002), the real distinction to be made between types of learning is not necessarily related to the setting in which they take place (i.e. formal learning in schools as opposed to informal learning in national parks or museums), but is based on the underlying motivation and interests of individual learners. In this way, free-choice learning 'recognises the socially-constructed nature of learning – the interchange that goes on between the individual and his or her socio-cultural environments – since implicit in the construct of free-choice learning is the ability of the learner not only to choose what to learn, but also where and with whom' (Falk 2005: 272).

Despite the emergence of these more complex understandings of environmental learning, however, much of the existing research in environmental education and related fields still largely fails to address learning as a process which is neither simply imposed (by formal institutions and policies) nor the result of individual agency (as in the case of 'free-choice' learning), but is also part of lived experience and social interaction. As a result, much of the existing research continues to treat learning as a somewhat mechanistic process in which learners 'choose' whether or not to participate.

One notable exception is a small, but rapidly growing, literature on 'social learning' which explores the ways in which individuals and groups (including communities and even whole societies) can be involved in learning processes (cf. Wals 2007; Reid et al. 2008). This work is linked to research and understandings of sustainable development, and the potentially new approaches to teaching and learning that it requires. Harold Glasser (2007), for example, suggests that social learning can be both 'passive' and 'active'. While passive social learning indicates learning which results from observation or listening but which does not require direct interaction with others, active social learning takes place via processes of dialogue and 'has the potential to promote more open, equitable, and competent learning processes' (2007: 15). This body of work represents a relatively new area of research on environmental education, as well as an opportunity to make connections to thinking in mainstream education research.

All of these specialist areas of research also already resonate strongly with more general educational research on learning, which for the last several decades has illustrated the incredible complexity of learning processes (for an excellent review, see Illeris 2007). This work has argued that learning is best understood as a complicated process of acquisition, accommodation, interpretation and capacity change, and that it is influenced by a number of individual and social factors. The work of key theorists such as Vygotsky and Piaget, for instance, continues to influence educational research on the complex nature of learning and its connections to learning, education and social change (cf. Vygotsky 1934; Piaget 1981; Mezirow et al. 2000).

In broad terms, then, each of these approaches represent attempts to develop a more sophisticated understanding of how people learn about the world around them – both as part of developing better conceptual understandings as well as for the purposes of making learning more effective/efficient. This is perhaps especially true for those who see environmental learning as a route to promoting engagement with environmental management or sustainable development, and for those who argue that sustainable development is in and of itself fundamentally a process of learning (cf. Scott and Gough 2003; Sterling 2001). Here, I use these emerging understandings from research as a framework for exploring diverse approaches to, and perspectives on, environmental education and learning in Costa Rica.

For this analysis, I also draw on one particularly useful metaphor for understanding the inter-connections between different spaces for learning. It was originally proposed by educational researchers Mark St. John and Deborah Perry (1993) who, in their discussion of effective methods for the analysis and evaluation of science museum exhibitions, recommend a consideration of museums as part of the 'educational infrastructure' of a nation. They define 'infrastructure' as the structures, systems and conditions that provide support to a wide range of economic and social activities, and argue that:

> 'Just as the economic health of a nation depends on the strength of its infrastructure, so the scientific and educational literacy of the nation depends on its educational infrastructure. It is very important to note that the educational infrastructure is not only, or even primarily, made up of physical resources. Rather than being composed of bridges, highways and water systems, the educational infrastructure can be thought of as an interwoven network of educational, social and cultural resources.' (St. John and Perry 1993: 60)

An exploration of a nation's educational infrastructure allows for attention to be given to the many diverse individuals and institutions engaged in educational projects, the links between them, and the broader social and economic relationships in which they are located. In this way, it may be possible to explore – from a more holistic point of view – the ways in which diverse groups and individuals engage in environmental teaching and learning.

Addressing a particular 'educational infrastructure' in this way also reflects contemporary anthropological interest in redefining 'the field' (Gupta and Ferguson 1997), and a growing interest in multi-sited research (Marcus 1998). From the work of George Marcus, in particular, I have drawn on his multi-sited strategies for 'following connections, associations, and putative relationships' (1998: 81) – in this case the concept and content of environmental education – through multiple sites and levels of engagement. The more nuanced understanding of the linkages between knowledge and educational infrastructures which can result from such an approach is particularly important in the context of contemporary discussions of 'knowledge economies' in which 'learning represents a fundamental source of capital, perhaps even superseding the industrial revolution triumvirate of money, labour and land' (Falk 2005: 274). In the case of Costa Rica in particular, a strong argument can be made for the economic importance of particular kinds of knowledge – especially about the environment and conservation – both for individuals and for the nation as a whole.

Structure of the Book

To explore the tensions surrounding environmental learning in a range of educational settings, and also to emphasise the connections between them, the following chapters – including one on formal education, one on NGO programmes and a third on education in public spaces – have been written in such a way that the three central arguments outlined above flow from chapter to chapter. (For the busy reader with a particular interest in one of those areas, the chapters can also be read individually, although some of the linkages and background details may be lost in the process.) I have also attempted to keep the academic and technical jargon to a minimum, in the hopes of making the book accessible to as wide and diverse an audience as possible. For ease of reading, footnotes have also been limited, although some readers may find the notes about additional information and resources of interest.

Structure of the Book

The book begins with a look at Costa Rica's national environmental education 'infrastructure' and at the history of education and educational ideologies in the country, in order to highlight some of the reasons why environmental education has proved to be so popular with both policy makers and the public (Chap. 2). While in some parts of the world environmental education has been marginalized because of concerns over the radical challenge it poses for existing educational and economic systems, in Costa Rica it has found broad acceptance. This reflects a national emphasis on citizens' entitlement to education and on the role of education in the promotion of participation in democratic processes, as well as the state's strategic focus on the promotion of scientific research, knowledge and conservation. This strong support from the state, I argue, is key to the high level of environmental interest and awareness in the country, and acts as a buttress to both the continuance of such educational programming and the important debates surrounding it.

Although environmental education is strongly supported by the state and the general public in Costa Rica, however, the responsibility for its implementation in schools overwhelmingly falls to schools and classroom teachers. In Chap. 3, I provide an introduction to the community of Monteverde, outline environmental education efforts in the community's formal education sector, and argue that classroom teachers face a number ideological, institutional and economic constraints to the implementation of environmental education in practice. Using research material gathered from both state and private schools in Monteverde in 2002–2003, the chapter discusses these limits on implementation, as well as opportunities for innovation, from the perspectives of local educators.

In addition to programmes in the formal education sector, local conservation organisations also play an important role in provision of environmental education in Monteverde by providing support and additional programming for state and private schools (Chap. 4). In organising programmes, however, educators in these NGOs negotiate complicated relationships with diverse community members, and between powerful local, national and international interests. The protectionist forest conservation agendas promoted by many local organisations and the powerful influence of international scientific researchers within many NGOs, in particular, often have a significant impact on the nature of local environmental education programmes. Using case material gathered from working with two local conservation organisations during the fieldwork year, I outline the programmes of two local environmental educators, and explore both their differing perspectives on the role of environmental education in conservation and community development as well as the ways in which their work mirrored long-standing debates in research, policy and practice between natural science-based and socially-engaged approaches to environmental learning.

Finally, in Chap. 5, I outline how the community of Monteverde is organised around a multitude of local interest groups and organisations through which residents interact, gather information and make decisions about local concerns. In 2002–2003, these groups made concerted efforts to communicate with other members of the community and to provide spaces for discussion (through workshops, seminars, lectures, public consultations and educational displays) of local concerns. The chapter details the importance of these spaces in two ways.

Firstly, public spaces offered opportunities for learning about local environment and development concerns, and were therefore an integral part of the extensive local network of environmental learning. Secondly, community members also used these spaces to strategically employ their knowledge of local affairs and relationships in order to address issues occurring *outside* these sites and in the community more generally. In this ways, these sites served as important reflexive spaces where community members could meet, debate and negotiate local environmental management and community development decisions.

The book then concludes with a few reflections on the implications of this research for environmental education and related fields, and suggests some ways forward for future research and policy.

References

Arnove, R., Franz, S., Mollis, M., & Torres, C. A. (1999). Education in Latin America at the end of the 1990s. In R. Arnove & C. A. Torres (Eds.), *Comparative education: The dialectic of the global and the local* (pp. 305–328). Oxford: Rowman and Littlefield.

Bangay, C., & Blum, N. (2010). Education responses to climate change and quality: Two parts of the same agenda? *International Journal of Educational Development, 30*(4), 359–368.

Bernard, R. (1988). *Research methods in cultural anthropology*. London: Sage.

Blum, A. N. (2006). *The social shaping of environmental education: Policy and practice in Monteverde, Costa Rica*. Unpublished doctoral thesis, University of Sussex, Brighton, UK.

Bourn, D. (2008a). Development education: Towards a re-conceptualisation. *International Journal of Development Education and Global Learning, 1*(1), 5–22.

Bourn, D. (2008b). *Global skills*. London: Learning and Skills Improvement Service. http://www.lsis.org.uk/Libraries/Documents/GlobalSkills%20Nov08_WEB.sflb.

Boyte, H., & Evans, S. (1992). *Free spaces: The sources of democratic change in America*. Chicago: University of Chicago Press.

Boza, M. A., Jukofsky, D., & Wille, C. (1995). Costa Rica is a laboratory, not ecotopia. *Conservation Biology, 9*, 684–685.

Courtenay-Hall, P., & Rogers, L. (2002). Gaps in mind: Problems in environmental knowledge-behaviour modelling research. *Environmental Education Research, 8*(3), 284–297.

Cousins, L. H. (1999). 'Playing between classes': America's troubles with class, race, and gender in a black high school and community. *Anthropology and Education Quarterly, 30*(3), 294–316.

Dillon, J., Kelsey, E., & Duque-Aristizábal, A. M. (1999). Identity and culture: Theorising emergent environmentalism. *Environmental Education Research, 5*(4), 395–406.

Eddy, E. M. (1985). Theory, research and application in educational anthropology. *Anthropology and Education Quarterly, 16*(2), 83–104.

Evans, S. (1999). *The green republic: A conservation history of Costa Rica*. Austin: University of Texas Press.

Falk, J. H. (2005). Free-choice environmental learning: Framing the discussion. *Environmental Education Research, 11*(3), 265–280.

Falk, J., & Dierking, L. (2002). *Lessons without limit: How free-choice learning is transforming education*. Walnut Creek: AltaMira Press.

Figueres Olsen, J. M. (1996). Sustainable development: A new challenge for Costa Rica. *SAIS Review, 16*, 187–202.

Fine, M., & Weis, L. (1998). *The unknown city: Lives of poor and working class young adults*. Boston: Beacon.

References

Fine, M., Weis, L., Centrie, C., & Roberts, R. (2000). Educating beyond the borders of schooling. *Anthropology and Education Quarterly, 31*(2), 131–151.

Freire, P. (1972). *Pedagogy of the oppressed*. Middlesex: Penguin Books Ltd.

Glasser, H. (2007). Minding the gap: The role of social learning in linking our stated desire for a more sustainable world to our everyday actions and policies. In A. Wals (Ed.), *Social learning towards a sustainable world: Principles, perspectives and praxis* (pp. 35–61). Wageningen: Wageningen Academic Publishers.

González Gaudiano, E. (1999). Otra lectura a la historia de la educación ambiental en América Latina. *Tópicos en Educación Ambiental, 1*(1), 9–26.

González Gaudiano, E. (2007). Schooling and environment in Latin America in the third millennium. *Environmental Education Research, 13*(2), 155–169.

Gough, A. (1997). *Education and the environment: Policy, trends and the problems of marginalisation*. Melbourne: The Australian Council for Educational Research Ltd.

Gough, S., & Scott, W. (2007). *Higher education and sustainable development: Paradox and possibility*. London: Routledge.

Grillo, R. D., & Stirrat, R. L. (Eds.). (1997). *Discourses of development: Anthropological perspectives*. Oxford: Berg.

Gupta, A., & Ferguson, J. (Eds.). (1997). *Anthropological locations: Boundaries and grounds of a field science*. Berkeley: University of California Press.

Ham, S. (1992). *Environmental interpretation: A practical guide for people with big ideas and small budgets*. Golden: North American Press.

Hammersley, M., & Atkinson, P. (1995). *Ethnography: Principles in practice*. London: Routledge.

Heimlich, J., & Ardoin, N. (2008). Understanding behavior to understand behavior change: A literature review. *Environmental Education Research, 14*(3), 215–237.

Henze, R., & Davis, K. A. (1999). Authenticity and identity: Lessons from indigenous language education. *Anthropology and Education Quarterly, 30*(1), 3–21.

Hess, G. A. (1999). Keeping educational anthropology relevant: Asking good questions rather than trivial ones. *Anthropology and Education Quarterly, 30*(4), 404–412.

Honey, M. (1994). *Hostile acts: US policy in Costa Rica in the 1980s*. Gainesville: University Press of Florida.

Honey, M. (1999). *Ecotourism and sustainable development: Who owns paradise?* Washington, DC: Island Press.

Horvat, E. M., & Antonio, A. L. (1999). 'Hey, those shoes are out of uniform': African American girls in an elite high school and the importance of habitus. *Anthropology and Education Quarterly, 30*(3), 317–342.

Huckle, J., & Sterling, S. (Eds.). (1996). *Education for sustainability*. London: Earthscan.

Illeris, K. (2007). *How we learn: Learning and non-learning in school and beyond*. London: Routledge.

IUCN. (1980). *World conservation strategy*. Gland: IUCN/UNEP/WWF.

Jaritz, K. (1996). Environmental education in teacher training: A case study at seven German colleges and universities and its outcome. *Environmental Education Research, 2*(1), 51–62.

St John, M., & Perry, D. (1993). A framework for evaluation and research: Science, infrastructure and relationships. In S. Bicknell & G. Farmelo (Eds.), *Museum visitor studies in the 90s* (pp. 59–66). London: Science Museum.

Kola-Olusanya, A. (2005). Free-choice environmental education: Understanding where children learn outside of school. *Environmental Education Research, 11*(3), 297–307.

Kollmuss, A., & Agyeman, J. (2002). Mind the gap: Why do people act environmentally and what are the barriers to pro-environmental behaviour? *Environmental Education Research, 8*(3), 239–260.

Koran, J. J., Longino, S. J., & Schafer, L. D. (1983). A framework for conceptualizing research in natural history museums and science centers. *Journal of Research in Science Teaching, 20*(4), 325–339.

Kwan, T., & Lidstone, J. (1998). Understanding environmental education in the People's Republic of China: A national policy, locally interpreted. *Environmental Education Research, 4*(1), 87–98.

Lai, O. K. (1998). The perplexity of sponsored environmental education: A critical view on Hong Kong and its future. *Environmental Education Research, 4*(3), 269–284.

Lave, J., & Wenger, E. (1991). *Situated learning: Legitimate peripheral participation*. Cambridge: Cambridge University Press.

Lotz-Sisitka, H. (2004). Environmental education research and social change: Southern African perspectives. *Environmental Education Research, 10*(3), 291–295.

Marcus, G. (1998). *Ethnography through thick and thin*. Princeton: Princeton University Press.

McKeown, R., & Hopkins, C. (2003). EE ≠ ESD: Defusing the worry. *Environmental Education Research, 9*(1), 117–128.

Mezirow, J., & Associates (Eds.) (2000). *Learning as transformation: Critical perspectives on a theory in progress*. San Francisco: Jossey-Bass.

Morales-Gómez, D. A., & Torres, C. A. (Eds.). (1992). *Education, policy and social change: Experiences from Latin America*. London: Praeger.

Ouyang, H. (2000). One-way ticket: A story of an innovative teacher in China. *Anthropology and Education Quarterly, 31*(4), 397–425.

Palmer, J. (1998). *Environmental education in the 21st century: Theory, practice, progress and promise*. London: Routledge.

Palmer, J., Suggate, J., Robottom, I., & Hart, P. (1999). Significant life experiences and formative influences on the development of adults' environmental awareness in the UK, Australia and Canada. *Environmental Education Research, 5*(2), 181–200.

Pellegrini Blanco, N. (2002). An educational strategy for the environment in the national park system of Venezuela. *Environmental Education Research, 8*(4), 463–473.

Piaget, J. (1981). *Intelligence and affectivity: Their relationship during child development*. Palo Alto: Annual Reviews, Inc.

Pike, G. (2008). Global education. In J. Arthur, I. Davies, & C. Hahn (Eds.), *The Sage handbook of education for citizenship and democracy* (pp. 468–480). London: Sage.

Rahnema, M., & Bawtree, V. (Eds.). (1997). *The post-development reader*. London: Zed Books.

Reid, A., Jensen, B., Nikel, J., & Simovska, V. (Eds.). (2008). *Participation and learning: Perspectives on education and the environment, health and sustainability*. Dordrecht: Springer.

Rickinson, M. (2001). Learners and learning in environmental education: A critical review of the evidence. *Environmental Education Research, 7*(3), 207–320.

Rickinson, M., Lundholm, C., & Hopwood, N. (2009). *Environmental learning: Insights from research into the student experience*. Dordrecht: Springer.

Sachs, W. (Ed.). (1999). *The development dictionary: A guide to knowledge as power*. London: Zed Books.

Sauvé, L. (1996). Environmental education and sustainable development: A further appraisal. *Canadian Journal of Environmental Education, 1*, 7–33.

Scott, W., & Gough, S. (2003). *Sustainable development and learning: Framing the issues*. London: Routledge.

Singleton, J. (1967). *Nich : A Japanese school*. New York: Holt, Rinehart and Winston.

Smyth, J. C. (1995). Environment and education: A view of a changing scene. *Environmental Education Research, 1*(1), 3–20.

Spindler, G. (1955). *Education and anthropology*. Stanford: Stanford University Press.

Sterling, S. (2001). *Sustainable education: Re-visioning learning and change* (Schumacher Briefing, Vol. 6). Totnes: Green Books Ltd.

Stocker, K. (2005). *'I won't stay Indian, I'll keep studying': Race, place, and discrimination in a Costa Rican high school*. Boulder: University Press of Colorado.

Street, B. (Ed.). (2001). *Literacy and development: Ethnographic perspectives*. London: Routledge.

Stromquist, N. P. (Ed.). (1992). *Women and education in Latin America: Knowledge, power, and change*. Boulder: Lynne Rienner Publishers.

Teitelbaum, A. (1978). *El papel de la educación ambiental en América Latina*. Paris: UNESCO.

UNCED. (1992). *Agenda 21, the United Nations programme of action from Rio*. New York: United Nations.

References

UNESCO. (1977). First intergovernmental conference on environmental education (Final Report). Tbilisi, USSR. Paris: UNESCO.

UNESCO-UNEP. (1976). The Belgrade Charter. *Connect 1*(1). Paris: UNESCO.

Vygotsky, L. (1934). *Thought and language*. Cambridge: MIT Press.

Wals, A. (1996). Back-alley sustainability and the role of environmental education. *Local Environment, 1*(3), 299–316.

Wals, A. (Ed.). (2007). *Social learning towards a sustainable world: Principles, perspectives and praxis*. Wageningen: Wageningen Academic Publishers.

Wax, M., Diamond, S., & Gearing, F. O. (1971). *Anthropological perspectives on education*. New York: Basic Books.

Williams, R. (1961). *The long revolution*. London: Chatto & Windus.

Willis, P. (1966). *Learning to labour: How working class kids get working class jobs*. Aldershot: Gower Publishing Company.

Chapter 2
Education, Environment, Development and the Costa Rican State[*]

Abstract Environmental education is often claimed to be at the centre of efforts to achieve sustainable development. Since the 1980s, Costa Rica has been one of the acknowledged international leaders in efforts to promote environmental learning, and national policy includes a three-fold national development strategy which simultaneously promotes education, conservation and ecotourism. As of yet, however, what is happening 'on the ground' has not been examined in much detail. This chapter addresses this gap in the literature by providing an overview of the diverse programmes and actors involved in environmental education in Costa Rica, as well as analysing the politics which surround its implementation.

Keywords Environmental education • Conservation • Discourse • National government • Non-governmental organisations • Policy

In many parts of the world, implementation of environmental education has proven difficult both because of economic or infrastructure constraints, and also because of the perceived 'radical challenge' environmental education poses to education (cf. Sterling 2001; Barraza et al. 2003). In general terms, this challenge revolves around the promotion of environmental education as a potential tool for social change. Although international policy statements which clearly support this aim – such as those found in *Agenda 21* – are commonly endorsed by policy makers in international meetings and conferences, the implementation of environmental education in many national contexts has often proven complicated[1]:

> '…. if we do accept a more socially analytical approach to environmental education, this has its own problems within the formal education system. The overt or critical stance to

---

[*]A version of this chapter was previously published as Blum (2008).
[1]There is, for example, an extensive literature on these issues in the UK (cf. Huckle and Sterling 1996; Sterling 2001; McKeown and Hopkins 2003; Lavery and Smyth 2003).

N. Blum, *Education, Community Engagement and Sustainable Development: Negotiating Environmental Knowledge in Monteverde, Costa Rica*, DOI 10.1007/978-94-007-2527-0_2, © Springer Science+Business Media B.V. 2012

social values and ways of life can create concerns about motivations, objectivity and sometimes relevance to what most teachers think they ought to be doing. Using education to challenge, even alter, social attitudes and values and thereby socio-economic systems – particularly if these are preconceived – poses enormous issues of acceptability from teachers, school managers, parents and the local and central government.' (Martin 1996: 46)

In Costa Rica, however, discourses of the role of environmental education in social change appear to have meshed relatively easily into existing national ideologies of education. These discourses arose out of the particular social and political history of the nation, beginning just after independence in the nineteenth century, and argue that education is an entitlement of all citizens, is essential for the promotion of participation in democratic governance, and is the most important means of promoting the development of both Costa Rican society and of individual citizens (cf. Fischel Volio 1992; Pérez 1981). More recently, national policy discourses related to environmental education have claimed that it is the 'ideal instrument' for building an environmentally conscious society and stimulating social change towards sustainable development (MEP 2002).

The central location of education within political, social and economic discourses in the Costa Rican context can not be underestimated. In the words of political scientist John Booth, education in Costa Rica has been granted the status of 'a virtual civil religion, embraced by rulers and citizens alike' (1998: 94). Between 1900 and 1950, state investment in formal education represented approximately 16% of the national budget, and had increased to almost 30% by the 1970s (Booth 1998: 94). The expansive infrastructure that developed through this intensive investment is commonly believed to be the reason for the nation's high levels of social development. In 2003, Costa Rica ranked significantly higher in terms of adult literacy (95.8% of the population over the age 15) and human development (ranked 47th on the human development index) than any of its Central American neighbours (UNDP 2005: 219). Historically strong connections between state investment and education have led to high expectations: 'Such great national investments in education and the national myths about its value have created powerful vested interests and expectations. Citizens demand education services from the government. Rural communities want neighborhood schools, even though tiny rural schools may deliver inferior education' (Booth 1998: 94). National newspapers and media further stimulate these discussions by offering frequent commentary on the condition of the national educational system.

The Costa Rican state has also long been praised for its green policies and progressive social programmes, and is an active and visible participant in international environmental policy-making. The current national constitution, for example, includes an article [#50] which provides for the right of every citizen to a 'healthy environment and a balanced ecology'. Costa Rica is also a major international site for natural science research in forestry, ecology and biotechnology, all of which bring in significant amounts of foreign investment from both non-profit and for-profit sectors. Since the mid-1990s, it has also been widely hailed as a conservation success story and a model for other nations to follow: 'The Costa Rican case has significance that transcends its small size. Because of the country's special attributes – democratic stability, an educated and environmentally aware citizenry, and a more egalitarian culture than most – Costa Rica provides a "best-case scenario" for forest preservation' (Brockett and Gottfried 2002: 8).

Indeed, research suggests that the general public in Costa Rica is keenly aware and informed about environmental issues and able to engage actively in debate about them. One national survey, conducted in 2002 by UNIMER Research International in conjunction with the national newspaper *La Nación*, reported that the majority of those surveyed believed that environmental degradation was among the five most significant problems faced by the nation, along with unemployment, violence, poverty and the high cost of living (UNIMER 2002). The survey also showed that the majority of respondents saw environmental education as essential to successful environmental conservation (Proyecto Estado de la Nación 2004: 30).

The creation of this 'environmentally aware citizenry' is often credited to the state's progressive environmental policy and promotion of public education programmes. Since the early 1970s, environmental education programmes have been implemented in the state education system, as well as throughout the extensive national system of protected areas. Statements about the need to educate the public about environmental issues are frequently placed at the centre of national legislation and policy. Policy-makers, educators, and conservationists throughout the country commonly link the provision of such environmental education programs with wider efforts in support of national conservation strategies and the expansion of the ecotourism industry. During the time of this research, discussions surrounding this three-fold approach to sustainable national development – and critiques of its implementation – could be found throughout the large grey literature authored by government agencies, non-governmental organisations, national university academic studies, and private business interests (e.g. MINAE 2000; Proyecto Estado de la Nación 2002).

Costa Rica is a small nation, both geographically (with a territory of only 19,600 square miles) and in terms of population (calculated at just over four million in 2003; INEC 2004). It is also often labelled an anomaly in Central America because of its long history of stable democratic governance and the famous abolishment of its national military in 1948 (cf. Fischel Volio 1992; Edelman and Kenen 1989). During the 1980s, when much of the rest of the region was plagued by civil wars and ethnic conflict, Costa Rican political leaders took an active role in re-establishing peace within the region, and President Oscar Arias (1986–1990; and also 2006–2010) was awarded the Nobel Peace Prize in 1987 for his efforts. Around the same time, former President Rodrigo Carazo Odio (1978–1982) famously claimed:

'The fundamental difference between Costa Rica and other Latin American countries is that Costa Ricans have cultivated a civilized spirit, a spirit antithetical to militarization and violence, capable of finding peaceful solutions to conflicts, and respectful of the rights of others. This respect has survived and flourished for two reasons: First, because education has fostered such an attitude; and second, because in the absence of weapons with which to impose an idea, the only weapon left is reason.' (cited in Reding 1986: 332)

With this reputation for peaceful, democratic governance and a high standard of living, the nation has regularly been offered up by its national political figures as an example for neighbours in Central America and beyond. Interestingly, Costa Rica only began to receive recognition for its conservation efforts in the late 1980s, following a decade in which the nation suffered from a severe economic crisis, high levels of foreign debt, the impacts of structural adjustment, and one of the highest deforestation rates in the world. So how can these oppositions be reconciled? Government officials, NGO campaigners, educators, students and ordinary citizens will tell you that it is because 'Costa Rica has more teachers than soldiers'.

The national commitment to education is a particular source of national pride, and when paired with the expansive network of national parks and private protected areas, it also constitutes an important economic resource. Scores of foreign researchers, tourists, conservationists, artists and NGO campaigners visit each year in the hopes of having a learning experience in the nation's biodiverse forests or along its miles of beaches. Since the early 1990s, this educational- and eco-tourism has been one of the nation's highest foreign capital earners, and the national government, international and domestic NGOs, and businesses have invested heavily in its continued success.

At the same time, although there is widespread public agreement about the necessity to provide education about environmental and development issues for the nation's children, young people and adults, what is less well understood is how these educational programmes contribute to environmental protection and sustainable development in practice. There are also significant and on-going tensions between educators, policy-makers, conservationists, researchers and the business community about what the most appropriate content and aims of these programmes should be. These diverse perspectives are rooted in a range of individual and organisational ideas and beliefs, and are also heavily influenced by complex social, historical and economic relationships at the local, national and international levels.

Before exploring local-level negotiations in Monteverde in depth, it is important to first look at the Costa Rican state's role in education provision and environmental management. This is because of the overwhelming size, visibility, and power of the state bureaucracy in both areas. As some commentators have strikingly noted, for instance, the state 'makes itself felt at every turn, not in the manner of a police state but as a bureaucratic giant that must be dealt with in order to own and drive a car, cut a tree, leave the country, build a shed, bury a body, buy or sell, employ or be employed' (Biesanz et al. 1999: 69).

The state's relationships to non-state organisations – both domestic and international – have also had significant impacts on national affairs and state administration. This is particularly the case in terms of environmental protection and advocacy, which are the focus of a much of the work conducted by non-governmental organisations (NGOs) across the country. Such is the influence of these non-state actors that some analysts have even labelled them a 'parallel state' (O'Brien 1997: 178). In addition to extensive international support for conservation, powerful international institutions – especially international donor agencies such as the World Bank – have also been heavily involved in the organisation and funding of the Costa Rican education system

(Heyneman 2003). These relationships have all, in turn, had a marked influence on state-organised environmental education programmes, both in terms of programme content and available resources.

Historical Development of National Education Discourses and the State Education System

The creation of national discourses of education has – as in many other post-colonial nations – been the result of a complicated exchange of ideas, philosophies and institutions. The Costa Rican state's involvement in formal education began shortly after achieving independence in 1821, under the influence of a small, influential group of liberal politicians who viewed education as the key to the nation's modernization and development (Booth 1998: 93). Many of these leaders had previously worked as teachers or had received advanced training in European educational centres, and therefore considered education to be of great importance. Of the five political figures involved in the creation of the document now considered to be the first national constitution (known as the *Pacto Social Fundamental Interino de Costa Rica*), for example, four had worked as teachers (Fischel Volio 1987: 62). Legislation which explicitly mentions the state's role in education is recorded as early as 1825 (when municipalities were legally charged with forming primary schools), and this was followed by a huge collection of regulations and legal frameworks which further extended the state's role in education provision. By 1844, the national constitution described education as a right of all citizens and the responsibility of the state (Booth 1998: 93). The number of state-funded schools throughout the territory steadily increased during the 1850s and 1860s, but economic crisis in the early 1880s resulted in the closure of many of these schools and the only national university.[2]

The government of the time, under the presidency of Mauro Fernandéz, responded to the crisis with an additional series of educational reforms and in 1886 began to invest even more heavily in education (Booth 1998: 93). This devotion to educational investment in the early post-independence era led to gradual increases in school attendance and levels of literacy. Even so, formal education was generally limited to small-scale institutions run by local governments or the Catholic church, and largely remained accessible only to members of the elite classes. As a result, levels of poverty and illiteracy remained high throughout the nation, and were particularly severe in isolated rural areas. In addition, political conflict left the new nation in a relatively disorganised state throughout the nineteenth and early twentieth centuries, during which time national politics was shaped by the dominance of elite political leaders, conflict and militarism.

---

[2]This was the *Universidad de Santo Tomás*, founded in 1821 to provide secondary and professional education (Booth 1998: 93). Although officially a state institution, it was largely dominated by Catholic clergy until the reforms of the 1880s (Biesanz et al. 1999: 25).

National elites, often labelled the 'coffee aristocracy' because they were initially led by powerful coffee export interests, supported policies of agro-export liberalism, and frequently conspired with the military in order to overthrow or impose Presidents who would cooperate in keeping taxes low and controlling workers. Violent conflicts in other parts of Central America at this time centred around ideological conflicts between Liberal and Conservative political ideologies. In contrast, in Costa Rica the Conservative movement had largely disappeared by the late nineteenth century, and so national politics during this era were dominated by Liberal ideologies. Costa Rican Liberals particularly supported the formation of a secular state with strictly limited control, a legal framework for protection of essential civil liberties, and the imposition of limits on the social and economic power of the Catholic church – especially in terms of its control over education. They believed that education should be the sole responsibility of a secular state, and would in this way act as 'a mechanism for material improvement as well as a means of modelling citizens' (Fischel Volio 1987: 37). Under the influence of these Liberal nation-builders, therefore, the Costa Rican state's role in providing education was afforded considerable importance, and despite continuing civil and political unrest, successive governments continued to endorse it as the right of all citizens.

The motivation for Liberal nation-builders to promote education for the general population arose not only out of ideological commitments, but also in response to high levels of poverty at the time of independence (Fischel Volio 1987: 61). Liberal leaders believed that improved education provision would help to alleviate poverty, but even with the achievement of impressive improvements to the national education system by the end of the nineteenth century, severe social and economic inequalities remained. Political leaders continued to seek reform and to debate issues surrounding educational provision, but their efforts were frustrated by 'deficiencies and scarcity of teaching personnel, parents' reluctance to comply with educational demands, and inadequate curricula' (Fischel Volio 1992: 144).

Political conflicts also continued to characterise the political life of the nation throughout the early twentieth century, with the result that presidential administrations were relatively unstable and often short-lived. By 1906, two distinct political parties – both endorsing Liberal ideals – had been formally established: the National Union Party (PUN; *Partido Unión Nacional*) and the Republican Party. The 1940s, however, was perhaps the most significant decade in Costa Rican political history in that it saw both a further escalation of political conflict, as well as an eventual resolution.

In 1940, in the midst of a period of serious labour unrest and the growing popularity and influence of Communist labour unionism and the Catholic church, Coffee aristocrat and Republican Party candidate Rafael Angel Calderón Guardia won the presidency and instituted a large package of social reforms. Calderón Guardia received sharp criticism from members of his own party as well as coffee export allies, however, when he pushed a social security programme that protected workers (the first of its kind in the country) through congress during the first year of his presidency. In response, he formed a populist coalition that included an influential Catholic archbishop as well as the Communist Party and its unions (Booth 1998: 43). With the strong political support of the coalition, Calderón Guardia was able to authorize a further sweeping series of social reforms, including the passage of a labour code which recognised the right of workers to strike, established social guarantees such as a minimum wage, and created the University of Costa Rica (*Universidad de Costa Rica*).

The reform movement continued throughout the next administration, led by President Teodoro Picado Michalski – Calderón Guardia's vice-president and successor. The new administration, however, was marked by intimidation and violence as the Costa Rican Communist Party (*Partido Vanguardia Popular*) and unions sought to reinforce their influence. Picado lost the 1948 election to the PUN's Ulate Blanco, and attempted to nullify its results with the support of the newly-gained *Calderonista* majority in the congress. This catalysed a brief, but bloody, civil war which lasted just 6 weeks from 10 March to 28 April 1948.

The warring parties eventually negotiated a peaceful settlement, and José Figueres, the charismatic leader of the winning revolutionary faction – the National Liberation army – ruled for 18 months before turning the presidency over to Blanco in 1949 (Booth 1998: 47). Both Figueres' short-term of office and Blanco's administration were characterised by continuing social reform. During the first year in office, for example, Blanco's government oversaw the creation of a new constitution which, among other celebrated social reforms, eliminated the national military, extended suffrage to women, prohibited the re-election of Presidents, and established a Supreme Electoral Tribunal which would ensure clean elections in the future (Booth 1998: 48). Analysts agree that it has largely been upon the foundation of the reforms of this period that the state has maintained its stability and strength.

Following the establishment of this more stable democratic system, beginning in the 1950s the state increasingly turned its attention to the expansion and further improvement of education in an effort to provide both the state itself and the growing industrial sector with sufficient technical and managerial staff. Costa Rican educational researcher Louis Mirón argues that increasing access to education during this era forced traditional elites to concede to a more fluid social structure in which education became a new and critical path for upward mobility into the growing middle or professional classes (Mirón 1989: 148). Education thus became a key means of encouraging economic growth as well as providing opportunities for individual advancement.

Organisation of the National Education System

Legislation passed during this era is also the foundation for the current national education system. The first Education Law (*Ley Fundamental de Educación*), passed in 1957, set out the central objectives for education in the nation, and further legislation in 1965 (*Ley Orgánica del Ministerio de Educación Pública*) outlined the administrative structure of the Ministry of Education (*Ministerio de Educación Pública*; MEP) and its role within the national education system. In 1973, President José Figueres (re-elected 1953–1958 and again 1970–1974) and Minister of

Education Uladislao Gámez approved a far-reaching National Plan for Educational Development (*Plan Nacional de Desarrollo Educativo*) which sketched out a new structure for the national education system, mandated free and compulsory education through the ninth grade, instituted a system of informal education, and expanded the country's universities (Mirón 1989: 149).

The National Plan restructured the previously existing system by replacing the traditional 'primary' and 'secondary' divisions with four new 3-year 'cycles', and this structure continues to be in use today.[3] In contemporary Costa Rica, the first three of these cycles (1st–3rd grade, 4th–6th grade, and 7th–9th grade) are compulsory. During the first and second cycles (in effect, primary education) students are required to study Spanish, social studies, science, mathematics and agriculture. Depending on the availability of teachers, students may also have the opportunity to take elective classes in music, physical education, religion, home economics, industrial and visual arts, and English or French.

In the third cycle (7th–9th grade) students receive classes in Spanish, social studies, English, French, mathematics, science, visual arts, music, physical education, religion, and a choice of industrial arts or home economics. Courses in the fourth cycle (10th and 11th grades) build even further on those offered in the third cycle with the addition of classes in philosophy and psychology, and with science studies divided into separate classes in physics, chemistry and biology. This final 'diversified' cycle is not compulsory, so students have the choice of continuing with their studies or entering the work force at that point. Those who continue in formal education must complete the relevant academic coursework (outlined above) as well as the requirements for a specialisation. This can be either in an academic programme focusing on in-depth study of science or the humanities for a further 2 years, or a technical programme concentrating on industrial, business or agricultural topics for a further 3 years. Each institution at this level is designated as offering specialised 'academic', 'technical-professional', 'vocational', 'tourism' or 'ecotourism', or 'environmental studies' programmes. The governing board of each school is authorised to make decisions regarding the most appropriate programmes for local circumstances, and to apply for approval from the Ministry of Education. The subjects and programmes available to individual students therefore varies a great deal by location, often depending on the local economy (including the local employment market) and the availability of trained teachers. Students in the urban Central Valley,[4] for example, are much more likely to attend 'academic' schools than their rural counterparts (where local circumstances may call for a greater emphasis on job skills or technical training), and therefore also to have easier access to higher education.

---

[3] For further details, see the MEP website (http://www.mep.go.cr/index.aspx) or documentation by the *Organización de Estados Iberoamericanos para la Educación, la Ciencia y la Cultura* (OEI; http://www.oei.es/quipu/costarica/#sis).

[4] This is the commonly used name for the large urban region located in the middle of the country, and in which are sited the capital, San José, and the two other largest urban cities in the country (Heredia and Cartago, both of which are also provincial capitals).

Students wishing to enter higher education are expected to complete all of the Ministry's academic requirements and to pass a tough series of national examinations. Admission to state universities is extremely competitive and contingent on national exam results. State-funded higher education is provided by three national universities with campuses in the Central Valley region: the *Universidad de Costa Rica* (UCR), the *Universidad Nacional* (UNA), and *Instituto Tecnológico de Costa Rica* (ITCR). In addition, the state funds a distance university, the *Universidad Estatal a Distancia* (UNED) which was founded in the 1970s and is modelled on the UK's Open University and Spain's National University. UNED provides correspondence and tutorial courses for students who cannot afford to live in or travel to the capital to continue their studies. The considerable, and increasing, demand for spaces in state higher education institutions has also recently led to the proliferation of private universities in the Central Valley region which offer a variety of courses of study and are often easier to enter, although more expensive to attend (cf. Twombly 1997).

Implementation and Effectiveness

Despite the impressive size of the national education system and the number of opportunities it offers, educators and policy makers frequently expressed concern during the time of this research about the overall educational quality offered by the state system. Educators cited particular problems with inefficient administration, a lack of resources and training for teachers, and frequent changes to Ministry of Education policy. One central, and long-running, concern linked to these issues is the tendency of many schools and teachers to place emphasis on teaching through memorization and rote learning. As Humberto Pérez, a leading authority in Costa Rican education, has succinctly described:

> 'Learning to learn: this is rarely taught in our schools. We often see primary, secondary, or university students copying information from a notebook or an encyclopedia, only to repeat it later in an exam without analysis or question. Our education continues to be bookish and by rote. It is believed that to read books *about* biology and history is to study biology or history.' (Pérez 1987: 56; original emphasis; cited in Biesanz et al. 1999: 207)

Classroom teaching methods frequently centre around teachers writing information on a chalkboard and requiring students to copy it into notebooks for later memorisation; alternately, they may read aloud and ask students to repeat the recitation. Student boredom with such teaching methods is often cited as a reason for misbehaviour and a general lack of interest in studying, particularly within adolescent student populations.

Ministry of Education policy and teacher training programmes in the national universities actively encourage teachers to use more innovative teaching methods, but in practice many educators are limited in their classroom practice by the widespread lack of resources or appropriate training. Educators argue that teacher training programmes, for example, are not effectively co-ordinated between the Ministry and the state universities, leaving new teachers unprepared for the demands of the curriculum and the classroom (Fischel Volio in Morales-Gómez and Torres 1992: 151–152). Classroom teachers complain that they are expected to 'be creative' and at the same time to prepare students for the Ministry's national exams – which are centred around mastery of a particular set of facts.[5] Others comment that advice from the Ministry changes so often – generally when the leadership is replaced by each new presidential administration – that it is simply easier to continue using older (rote) teaching methods. Frequent budget shortfalls further exacerbate the existing scarcity of resources which educators – especially in poorer, rural areas – continually face.

Teachers' frustrations with the state system frequently result in strikes by the national teachers' unions. These strikes can paralyse the school system, as they did more than once during the time of this research. The most serious of these began in December 2002 when the *Sala IV* (the nation's constitutional court) ruled that the Ministry of Education had to abide by its earlier agreement to a Central American convention which required a minimum of 200 school days per year. The ruling was in response to an emergency measure proposed by the Ministry to shorten the upcoming school year to 174 days because it lacked the funds to cover teachers' salaries for the full 200 school days. The beginning of the school year was delayed for 3 weeks while government officials contested the court ruling, with classes finally beginning on 11 February 2003. Conflicts between the Ministry and the teaching unions continued, however, due to the Ministry's inability to pay the additional wages as well as over changes to the pension structure for state teachers, and many teachers continued to refuse to teach classes. Unspecified 'computer problems' at the Ministry shortly after the beginning of the new school year also resulted in incorrect or insufficient pay to more than 700 teachers across the country, which served to further enflame the conflict (Tico Times 2003). The state also failed to provide promised funds for meals for poor students in 3,804 school cafeterias across the country, and there were on-going problems with provision of student transportation (see La Nación 2003c). The then-Minister of Education, Astrid Fischel Volio, was forced to resign on 3 June 2003, but the strike lasted another 3 weeks until a compromise was finally reached to return to classes on 30 June 2003. By that time, students across the country had missed a total of 1 month of classes, only 11 days of which were eventually re-scheduled.

The Ministry frequently struggles with such large-scale shortages of funding partly because of internal organisational problems, but also due to the state's wider economic difficulties. The national economic crisis of the 1980s and the impacts of IMF-imposed structural adjustment programmes have had especially significant impacts on the state education infrastructure and the state's ability to provide educational services. The state's central role in education provision – through the Ministry

---

[5] Biesanz et al. (1999) cite example questions such as: 'Who was Gaspar's sweetheart?' (in a subplot of *Don Quixote)*, 'What are the Pope's parents' names?', or 'What was Braulio Carrillo's most identifiable characteristic on his strolls in San José' (Carrillo is a famous figure in the history of Costa Rica; the answer is 'his ebony cane') (1999: 207).

of Education and other agencies – has firmly tied the fortunes of the education system to the changing economic fortunes of the national economy.

Strong links between education and statecraft have therefore both imposed limits and presented opportunities for educational provision and innovation. In particular, while economic crisis has frequently posed limits on the ability of the state bureaucracy to provide educational services, the state's rapid assimilation of discourses of environmentalism and sustainable development has also attracted substantial international investment, aid, and other forms of funding for both conservation and education. The explicit links drawn by national leaders between the national education system and the possible future successes of sustainable development efforts have provided powerful avenues through which the state can communicate with, and benefit from, relationships with other nations and international organisations. In these ways, environmental education has become both ideologically and economically important to the nation as a whole, in both domestic and international spheres.

State Environmental Management, the Ecotourism Industry and Scientific Research

Despite the high profile nature of the concept of sustainable development in Costa Rica, state-funded environmental protection programmes have not always been entirely successful in practice. During the 1980s, for example, the nation was losing an estimated 4% of its forest cover annually – the highest deforestation rate in the western hemisphere at the time (Carriere 1991: 188). Between 1970 and 1980 more than 7,000 km² were cleared, and by 1987 total forest cover had been reduced to only 31% of the available land mass or approximately 16,000 km² (Carriere 1991: 188). Since that time, analysts have blamed these high deforestation rates on large commercial interests such as logging, mining and cattle production (cf. Carriere 1991; Edelman 1995), as well as on illegal squatting and land clearance enabled by the state's policy of encouraging settlement of the interior in the nation's early statehood (Augelli 1987). Perhaps most worryingly for the state and the national economy, this deforestation brought with it serious concerns about soil erosion and land degradation during a period of explosive population growth.

A National Forestry Law (*Ley Forestal*) had in fact been passed in 1969, and was intended to address the issue of deforestation long before it reached the crisis levels of the 1980s.[6] The legislation established a legal and administrative structure to designate and administer a system of protected areas in which logging and agricultural activities would be strictly prohibited. Unfortunately, the effectiveness of the law was hampered by a lack of long range planning, ineffective permitting procedures

---

[6] This was not actually the first environmental legislation in the country, but it is commonly referred to by observers inside and outside of the country as the beginning of modern environmental legislation. For more on environmental management and legislation prior to this, see Fournier (1991) and Evans (1999).

and a general lack of funds and trained personnel, with the result that deforestation rates continued to rise. In 1971, forestry biologist and co-founder of the Tropical Science Centre, Joseph Tosi, famously predicted that if deforestation continued unabated, Costa Rica would have virtually no forested areas left by 1985 (Evans 1999: 50). Faced with such dire predictions and in response to calls from conservationists and researchers both inside and outside of the country, the state began to pass stricter legislation and to devote more resources to forest protection.

Despite the subsequent establishment of a huge body of environmental legislation and a large and powerful government bureaucracy to administer it, however, during the time of this research the state still faced serious problems with inefficient administration and ineffective enforcement. This was a problem not only in terms of environmental legislation and management, but throughout the national political arena. Costa Rican historians, political scientists, and other observers have blamed this widespread disjuncture between political rhetoric and lived reality on the recognised tendency of political leaders to avoid conflict:

> 'Consensus is often achieved at the expense of decisiveness… Costa Rica is often called a nation of laws and lawyers…. But many laws, unsupported by a realistic plan or by resources for enforcement are simply evidence of good intentions… Symbolic solutions satisfy the formalistic, legalistic outlook common among Costa Rican leaders. They meet to discuss a problem in committees, seminars, and workshops; proclaim the correct solution; pass a law or create a new agency; and presto! The problem is considered solved.' (Biesanz et al. 1999: 77)

Such strategies may allow leaders to avoid conflict – in national parlance to *quedar bien* – but they often seem to inhibit real problem-solving.[7] Furthermore, economic constraints often make it difficult for state agencies to enforce the continually growing body of environmental legislation or to manage any new initiatives, regardless of how important they are deemed to be in national discourse. For example, national media regularly noted in 2003 that the Ministry of Environment and Energy (*Ministerio del Ambiente y Energía*; MINAE)[8] – the only government agency with the legal authority to administer all environmental legislation, policies, and programmes – suffered from a continual lack of funding and staff (cf. La Nación 2003a, b). The state also continues to struggle with a large amount of foreign debt, and multinational corporations (mostly based in the US) have begun to dominate many domestic markets in the last two decades, often at the cost of smaller Costa Rican businesses. Despite strong public and private support for conservation and scientific research in the country, the state therefore constantly struggles to both ensure national economic stability and to administer environmental legislation and policy, including environmental education.

At the same time, it is the nation's long-standing emphasis on maintaining peaceful relationships – both between individuals and with other nations – that is at the root

---

[7] The concept of *quedar bien* is multi-faceted, and defies simple translation. Literally, it means 'to maintain good relationships', but it is more generally used to describe the effort to make a good impression, to avoid conflict, and to appear polite and amiable in social relationships.

[8] More recently renamed the Ministry of Environment, Energy and Telecommunications (Ministerio del Ambiente, Energía y Telecomunicaciones; MINAET).

of its international reputation for peace and stable democratic governance. This reputation has played a significant role in the state's successful international promotion of tourism, alongside forceful marketing of the territory's great natural beauty, favourable climate and beaches. According to the Costa Rican Tourism Board (*Instituto Costarricense de Turismo*; ICT) – the state agency responsible for national tourism development – more than one million tourists were visiting the country annually during the time of this research (ICT 2003). More recently, this has doubled to an estimated two million visitors per year (Benavides Jiménez 2009). The national system of protected areas encloses an estimated 30% of the national territory, and was the focus of the tourism industry throughout the 1970s (Evans 1999; Wallace 1992; Boza 1993). In the early 1990s, the ICT initiated marketing campaigns specifically to attract high price, low impact 'ecotourism' to the country.[9] Since then tourism has clearly continued to grow, with much of its popularity based on the nation's reputation for well-preserved tropical ecosystems and opportunities for 'rainforest adventures'. Integral to ecotourism are minimisation of the environmental and social consequences of tourism, and the inclusion of education and awareness-raising activities. Eco-tour companies in Costa Rica thus seek to provide travellers with experiences of the natural world as well as information about conservation activities and environmental issues in the areas they visit.

This shift to an emphasis on tourism development was a significant move away from a historically agrarian economy based on coffee, beef and banana production. After a slow start, the tourism industry experienced a boom in 1987 and annual visitation rates have continued to grow ever since. Statistics provided by the ICT suggest that international visitation is highest during the dry season (December to March), followed by a significant drop in numbers, then a second short spike during the 'shoulder season' (July and August, concurrent with school holidays in the United States), and a gradual build-up to the end of the year. In the 1970s, the majority of these international visitors came from other Central American countries, but regional political instability throughout the 1980s paired with steadily increasing rates of visitation from North America and Europe have since resulted in a shift towards predominantly North American visitation.

Travel to Costa Rica by North Americans and Europeans actually began early in the nation's history with visits by foreign scientists and naturalists. Scientific research in the country began as early as the 1840s when US, Danish and German scientists arrived to assess the territory's natural resources (Eakin 1999: 127). As part of their support for the establishment of the national education system, Costa Rican political leaders at the time were particularly interested in providing education and training in science and technology, and an influential group of European academics were subsequently recruited by the state to set up research institutes and

---

[9] According to the Ecotourism Society, the concept of ecotourism denotes, 'Responsible travel to natural areas that conserves the environment and improves the well-being of local people'. Also see Honey (1999).

provide training to Costa Rican citizens. National leaders believed that once a new generation of trained professionals could be established, the nation would begin to benefit from new technologies such as railroads, telegraphs, electricity and steamships (Eakin 1999: 128), and despite sometimes severe economic constraints the state continued to support scientific research (a significant portion of which was related to the development of coffee and banana production) throughout the end of the nineteenth and early twentieth centuries.

The early influence of European and North American scientists on the young country proved significant to the development of Costa Rica as an internationally-important site for research. Researchers such as Karl Hoffmann, Alexander von Frantzius, Henri Pitter, Pablo Biolley, Julian Carmiol, Gustavo Michaud, and Juan Rudin, were initially attracted to the region by utopian travel accounts and the growing literature on Costa Rican flora and fauna, but later settled in the country and were responsible for the establishment of important national scientific institutions (including the *Instituto Físicio Geographico*, the *Sociedad Nacional de Agricultura*, the *Observatorio Nacional*, and the *Museo Nacional*) (Janzen 1983: 2–4). Alexander von Frantzius' Costa Rican apprentice, José C. Zeledón, was also responsible for establishing the first links with the Smithsonian Institution, after which point 'the flow of United States researchers has never stopped' (Janzen 1983: 4). Well-known American researchers who have lived and worked in Costa Rica more recently, and who have played an influential role in the development of the national system of conservation areas include Archie Carr (established the Caribbean Conservation Corporation in Tortuguero in 1959 to protect and study sea turtle nesting grounds), Daniel Janzen (influential in the establishment and protection of Santa Rosa National Park, one of only two areas of protected tropical dry forests in the world), and Leslie Holdridge (co-founder of the Tropical Science Center; he created a classification system for tropical forests which is still in use today, see Holdridge 1947, 1967).

The state's heavy promotion of conservation, ecotourism and scientific research for the last several decades has continued to encourage strong links with international conservation and scientific research organisations, foreign universities, and individual researchers. These connections have provided significant financial and social benefits to the country, especially in terms of conservation programmes. While the state owns the majority of protected areas within the territory, by 2003 private organisations and individual owners protected an additional 1% of the territory (SINAC-MINAE 2003: 3) and were often more popular tourism destinations than state-owned areas. The size of these private land holdings varies widely, from large areas covering thousands of hectares to small-scale projects aimed at protecting community forests or watersheds. Both large and small scale private conservation projects frequently receive financial and administrative support from international organisations such as The Nature Conservancy, Conservation International and the World Wildlife Fund, among others. These large organisations are based in the Central Valley, and are heavily involved in administering international-level projects or running campaigns aimed at national policy makers (e.g. for sustainable timber certification or carbon trading initiatives).

Perhaps largely as a result of these international relationships, European (Enlightenment-influenced) perspectives on the natural sciences and research are a noticeable element of national legislation and discourses of development and environmental management in Costa Rica. Scientific understandings and analyses of national environmental issues are often heavily privileged over other ways of 'knowing'.[10] This is especially true in local contexts where scientific research has been the catalyst for the establishment of protected areas which, in turn, act as an economic base for neighbouring communities. In many important ways, the continuing emphasis on perspectives from the natural sciences and protectionist conservation schemes has had powerful economic and political implications for the nation. During the 1960s and 1970s, it attracted the attention of international and regional environmental NGOs and was one of the main reasons behind the creation of numerous domestic NGOs. All of these groups have since given significant financial and rhetorical support to the state's efforts to promote science and environmental learning to various sectors of the public (Fournier 1991: 79). In addition to attracting funding for conservation and research, the state has also negotiated a number of profitable biotechnology agreements, including a deal with US-based Merck Pharmaceuticals (Coughlin 1993) and a variety of international climate cooperation projects (Dutschke and Michaelowa 2000).

Discourses arising out of these natural science perspectives tend to support the strict protection of fragile ecosystems for scientific study, and as such they signify the continuance of what has been called the 'traditional conservation narrative' in Costa Rica. This traditional narrative emphasises the threat to wildlife populations – especially in developing countries – posed by direct human exploitation or the indirect impacts of population growth and demands for development, and advocates the imposition of strict protection regimes enforced by state authorities (Campbell 2002: 29–30). This discourse emerged in Costa Rica in the early twentieth century and is the foundation for the current national system of protected areas. A contrasting 'conservation counter-narrative' emerged in the late 1980s and early 1990s in conjunction with the rise of ecotourism and international attention to sustainable development. This narrative centres on 'sustainable use' schemes and advocates 'community-based conservation' that allows local populations to participate in conservation projects (ibid: 30–31). Evidence of contemporary negotiations between (and strategic use of) these two narratives in Costa Rica can be readily found within the national policy literature, as well as in documents published by a variety of conservation organisations, community or citizen's groups, and business interests. In local contexts, individuals with diverse interests such as farmers, landowners and conservationists, are also likely to call upon these discourses to legitimate their calls for action, and may collaborate with groups with similar interests in order to achieve common goals (cf. Nygren 1998).

---

[10] The nation's 21 recognised indigenous groups, for example, are both geographically isolated and overwhelmingly marginalized in national politics, although they have received some attention from international environmental organisations (cf. Mayorga et al. 2004; Vargas et al. 1999: 135).

State-Funded Environmental Education

State funding is provided for the promotion of environmental education in both formal and informal settings. The majority of state formal education sector programming (i.e. in schools) is administered by the Ministry of Education, but the national universities also provide opportunities for environmental learning. In addition, programmes in less formal educational settings such as national parks and other protected areas are organised under the auspices of the Ministry of Environment and Energy.

Schools

In the state-funded formal education sector, environmental education topics first began to be introduced in the Costa Rican national curriculum in 1977. The first national 'Environmental Education Master Plan' was published in 1987, amid growing recognition of the need for a department dedicated specifically to environmental education. In 1993, the Office of Environmental Education (*Oficina de Educación Ambiental*; OEA) – a division of the Ministry of Education – was formally established by executive decree (OEA 2002: 10). It is a relatively small division, but with large responsibilities. During the time of this research, it had a staff of five employees who were responsible for providing training on the environmental education requirements of the national curriculum to all of the nation's state school teachers in more than 1,600 state-funded schools. Between 1995 and 1999, the Office co-ordinated programmes which addressed issues such as forest conservation, sustainable watershed management, population growth, solid waste management, and energy conservation. It also organised environmental clubs in schools nationwide, and co-ordinated with a range of state agencies, non-state institutions, and international partners on initiatives related to environmental issues (OEA 2002: 11).

In 2003, a large portion of the Office's resources were being dedicated to training teachers on the national curriculum's environmental education requirements. Despite the major role that the state universities have played in national conservation efforts since the early twentieth century (Evans 1999: 21–23), many educators claim that teacher training programmes do not provide sufficient training in environmental education topics and teaching strategies. Teaching about environmental issues at the state universities has tended instead to be limited to other disciplines or areas of action. As early as 1975, for example, the *Universidad Nacional* established a School of Environmental Sciences which included an environmental education programme, in 1977 the *Universidad Estatal a Distancia* created a Centre for Environmental Education, and in 1994, the National Council of Vice-Chancellors created an Inter-University Commission for Environmental Education (CIEA) which works to 'environmentalise' (*ambientalizar*) all of the state universities (OEA 2002: 10). Co-ordination between education departments in the state universities – which are responsible for managing teacher training programmes – and the Office of Environmental Education, on the other hand, is minimal. This has meant that new teachers are largely unprepared to meet national curriculum requirements for environmental education in their classrooms.

In response to this need, the Office organised an extensive series of workshops and seminars for educators. Because of its limited time and financial resources, however, it is often difficult for the Office's staff to reach isolated (usually rural) schools and teachers. In 1999, the Ministry and the Office began attempting to address this problem by publishing self-training guides for teachers who did not have access to face-to-face training programmes. Each of these books contains lessons and self-guided activities that cover a particular area – 'Development in Harmony with Nature' (MEP and OEA 2002a), 'Human Intervention in the Biosphere' (MEP and OEA 2002b), 'Education for the Prevention of Disasters' (UNESCO and RNTC 2000), and 'Environmental Education Pedagogy' (MEP and OEA 2002c). The materials were designed so that teachers would be individually responsible for working through the guides and sending reports to their regional Ministry of Education representative, who could then certify their completion of the training.

During the time of this research, the Office was also heavily involved in a broader movement within the Ministry to promote the application of 'transversal themes' (*temas transversales*) as part of the national curriculum. These were initially identified by Ministry policy-makers in 2001 in consultation with representatives from the United Nations Population Fund.[11] The themes identified were 'human rights, democracy and peace', 'building a culture of environmentalism and sustainable development', 'health education', and 'sex education' (MEP 2002: 17). Rather than constituting a further curricular requirement, the themes were intended to cross-cut all other areas of the curriculum through integrated classroom activities. Among the many creative examples of this I heard during my fieldwork stay, one Ministry employee told me that he asked students in his music programmes to listen to 'natural music' (birds, wind, sea), and then engaged them in a discussion of what would happen to these sounds if forests are cut down or animals become extinct. Another Ministry employee suggested that: *Environmental themes can be incorporated into any subject. In mathematics, you can teach math skills like statistics by discussing population growth or changes in forest cover.*

In addition to the use of the transversal themes, older students in selected secondary schools also have opportunities to receive specialist training in environmental topics. In 2003 there were four secondary schools in the country that offered technical qualifications in 'ecotourism', as well as two others that offered an 'environmental studies' programme. This new environmental studies designation was a matter of some pride for the Office of Environmental Education. The schools offer both

---

[11] The idea of promoting specific topics or themes in a transversal manner originated as part of educational reforms in Spain in 1990, and is also often discussed in other Spanish-speaking countries (cf. Lencastre 2000; Luzzi 2000; Garcia Gómez 2000; González Gaudiano 2000; Reigota 2000; Roth 2000).

intensive studies of environmental issues as well as training in ecotourism and English. According to the head of the Office at the time, the new schools' explicit mission was to create nature lovers (*amantes de la naturaleza*) who will 'actively share their love of nature with others' after graduation. In concrete terms, this meant that the schools devoted a great deal more time and space within their specialist curriculum to environmental studies than any other state schools. Students in the 'environmental' schools received 20 h of environmental education instruction per week, for example, while their 'technical-professional' counterparts received only 6 h per week.

National Conservation Areas

Further environmental education in schools and within communities is provided by the Ministry of Environment and Energy through programmes in regional conservation areas. The Ministry was established in 1986, and is responsible for management of all public lands and all conservation issues aside from those relating to agricultural production. It is a large and powerful agency, which encompasses the Forestry Directorate, the Department of Wildlife, the Department of Geology and Mines, the National Zoo, and the National System of Conservation Areas (*Sistema Nacional de Areas de Conservacion*; SINAC). At its inception, the Ministry's strategy for national park management focused on strict protection – an approach which often marginalized neighbouring communities. The national parks were also initially administered as relatively independent entities, but in 1995 they were re-organised into the current National System of Conservation Areas. This change streamlined the government administration (state protected areas had previously been simultaneously administered by the forestry, wildlife *and* wild lands agencies), and also put in place regionally integrated management schemes as a response to the growing popularity of 'sustainable development' strategies. Ten 'conservation areas' were established across the country, each of which includes a core area for biodiversity conservation (usually a national park) and buffer zones for sustainable development activities such as controlled timber or firewood extraction, wildlife management and ecotourism. Although officially under the authority of the Ministry of Environment and Energy as well as SINAC, each of these conservation areas is relatively autonomously administered:

> 'The ten Conservation Areas have evolved into territorial units (state protected areas, private property, and urban zones), governed under one development and administrative strategy in which private, local, and federal management and conservation activities are interrelated, and solutions based on sustainable development are sought jointly with the civil society.' (Vaughn and Rodriguez 1997: 446)

As a product of this relative autonomy in management, each conservation area has its own management and development schemes (more recently, these have focused on sustainable development) which attempt to address the specific needs of each area and its inhabitants.

Environmental education programmes in the conservation areas are also relatively autonomously organised in order to suit particular local needs and conditions. In response to the wide diversity of programming which has resulted, in 1990 the Ministry of Environment and Energy established a national office for environmental education to co-ordinate amongst the disparate conservation areas. During the time of this research, the office was staffed by two national co-ordinators with responsibility for organising programmes, directing a national Environmental Education Commission composed of relevant ministers and area managers, and developing a national environmental education policy. According to one of the national co-ordinators I spoke to in 2003, however, not all of the conservation areas were actually able to run environmental education programmes at that time. Regions which received greater public attention, such as high density tourist destinations, were more likely to have the necessary resources and staff.

Among those that did offer programmes, there was a great deal of variation in terms of their content and approach. Some of these differences were based on the ecological character and specific conservation needs of individual areas (the educational content of projects organised in a cloud forest ecosystem is understandably different than for those located in coastal areas). Programme content and orientation also depended heavily on regional policy-makers and educators themselves. According to the national co-ordinator, the programming offered in Guanacaste province (in north-western Costa Rica), for example, focused largely on teaching the natural sciences, and the organisers specifically identified it as 'biological education', rather than environmental education. Programmes in the Osa Peninsula (on the southern Pacific coast), on the other hand, tended to be more strongly oriented towards teaching about local social issues and their connections to environmental management, as well as to more generally encouraging community involvement in protected area management decisions.

Defining and Implementing Environmental Education in the State System

Indeed, what fell under the umbrella of 'environmental education' in national, regional, and local discourses during the time of this research was often the subject of a fair amount of debate, not just in the Ministry of Environment and Energy, but also within and amongst the many other government agencies and non-governmental organisations. These debates centred on both the content as well as aims and goals of programming. More specifically, many national actors agreed that although the general population had achieved a significant level of 'environmental awareness' through the efforts of educators and environmentalists, there remained a great need to use education as a catalyst for action. As one environmental educator employed by the Ministry of Environment and Energy told me:

> *What has changed in Costa Rica is only the ideas and how much information people know. Everyone knows about environmental issues, but that isn't necessarily changing people's*

behaviour. They may know it's bad to dump a bunch of plastic bottles in the garbage or cut down a tree, but that doesn't mean they won't do it. We have increased our knowledge, but not how we are. The next step in education has to be to actually change behaviours.

A professor based at one of the national universities, agreed:

Environmental education is about teaching content to a certain extent, but it's not really environmental education unless it leads to action. What environmental education should do is not only change attitudes, but also behaviours, and it should empower people to make their own decisions. By showing people the consequences of certain behaviours, as well as giving alternatives, you can bring about change.

Underlying these discussions about stimulating change through environmental education are strong national discourses of education as an avenue for promoting particular, Costa Rican social values. The front cover of the Ministry of Education's 2002 'transversal themes' publication (MEP 2002), for example, was decorated with a piece of student artwork entitled 'Dialogue is the best way to achieve peace in every part of the world'.[12] The cover, designed by a group of students in the second cycle at a school in the Central Valley, shows a blue and green globe with six children dispersed along its surface. Each child carries a sign inscribed with a word: happiness, tolerance, hope, dialogue, love and peace. In the introduction to the publication, officials from the Ministry argue that environmental education is the 'ideal instrument' for building an environmentally conscious society and achieving sustainable development.

Ministry policies and publications regularly call upon the education community to teach learners about their interdependence on their biophysical, social, economic, political and cultural environments and to participate actively in the detection and solution of environmental problems in their local communities and the rest of the planet.[13] This emphasis on promotion of values through the educational process, argued the head of the Office of Environmental Education in 2003, is precisely why environmental education is an 'integrated' subject, rather than a single component of the national curriculum: *It isn't like other subjects. It is fundamentally about values, and changes in attitudes and aptitudes… environmental education should provide content – in terms of the curriculum – and also allow students to be reflexive about their own ideas and behaviours.*

Many educators argued, however, that implementation of these ideas could be difficult due to a number of more practical concerns. Classroom teachers frequently noted that the integrated style of teaching advocated by the Ministry and the Office of Environmental Education was difficult to implement due to both a lack of sufficient teacher training, and because of the heavy requirements of the national curriculum and the need to prepare students for content-based national exams. Environmental education (like the other transversal themes) was not examined via

---

[12] The original Spanish reads: '*En cualquier lugar del mundo, el diálogo es el mejor camino para alcanzar la Paz*'.

[13] For instance, the Ministry has also actively promoted a national values education programme for several years (cf. Martinez 1998; MEP 2004; Castillo 2006).

the state's national exams, and so was often neglected in classroom instruction in exchange for attention to basic subjects which are formally assessed. Educators also pointed to the frequent changes to the administrative structures of many government ministries as a limiting factor. These changes of personnel and policy tend to happen quite regularly – approximately every 4 years with the election of a new Presidential administration.[14] With each change the in-coming party installs a host of new Ministers, advisors, and other bureaucrats and creates a large volume of new policy and legislation. These 'new' policies are widely publicised as a credit to the new administration, but they often represent only slight substantive changes to previous ones. In practice they may also actually serve as an obstacle to the achievement of long-term goals by requiring expensive and time-consuming changes to administrative structures, organisation and bureaucratic processes.

The shifting and sometimes competitive character of the national bureaucracy also impacts upon individual actors within it. In the Ministry of Education this has implications for the fundamental conceptualisation of environmental education in policy, as well as the ways it is put into practice. According to one Ministry employee in 2003:

Since the 1970s, there have been continuous debates here between those who believe that environmental education should deal with the environment both in a biological sense and in terms of it social impacts, and those who think it should have a more limited scope. Often the limits people want to impose are mostly to do with them wanting to protect the areas in which they are already working.

Conflicts about the content and orientation of environmental education, as in the case of education provision in general, must therefore be seen as not simply linked to theoretical debates or arguments about cultural or social norms, but also to strategic decisions by actors involved in policy formation and implementation.

All of these discussions about the role of environmental education in building a sustainable society in Costa Rica also take place within a context of relatively scarce resources. This is because, despite its strong discursive commitment to environmental learning, the state has often been unable to provide sufficient financial support for programming. Scarce funding and resources are especially problematic at the level of the school and classroom. Many school buildings I visited in 2003 were in serious need of repair, and classrooms frequently lack basic supplies such as chalk, paper and textbooks. These issues can have a significant impact on the education system as a whole, but they also have particular impacts on environmental education because schools and educators who struggle with a lack of access to sufficient classroom materials and training are less likely to devote time and resources to 'special' (and un-assessed) parts of the curriculum. As a result, many students received little or no environmental education in practice, and schools often relied on outside

---

[14] Historically, the Presidency has tended to oscillate between the two main political parties – the 'social democratic' National Liberation Party (*Partido Liberación Nacional* or PLN) and the Social Christian Unity Party (*Partido Unidad Social Cristiana* or PUSC), a coalition of liberal opposition.

organisations such as conservation areas or NGOs – where they were available – to provide lessons and materials.

Concerns over limited resources are especially acute in rural areas. In response, both the state and non-state organisations have increasingly turned to mass media technologies as a means of communicating environmental messages and providing resources to large public audiences and to geographically isolated regions. During the time of this research, for instance, the Ministry of Education supplied a range of teaching and learning support for teachers and students in rural areas via serial radio programmes. Some of the topics included instruction in environmental education, mathematics, and English. The Ministry of Education was also increasingly involved in co-ordinating innovative training and education programmes using television broadcasts and internet technologies.

For its part, the Ministry of Environment and Energy has also utilised radio and television advertisements to build popular support for, and awareness of, the national park system and current conservation initiatives. In 2003, for example, a Ministry campaign urged citizens not to buy or procure wild birds for sale as pets. One television advertisement for the campaign began with a rather dramatic scene of a bulldozer crushing a pile of metal bird cages. This was followed by an emotional plea from a group of young people who asked members of the public to 'leave animals in the forest where they belong', adding that 'when your parrot dies, destroy the cage, so that one more bird can stay in the forest'.

Non-governmental organisations – including private protected areas and conservation organisations – have also used broadcast media to promote their programmes and opportunities for both leisure and learning activities. National media, including newspapers, television and radio broadcasts also routinely deal with issues of current environmental concern or promote participation in conservation projects. The popular daily newspaper, *La Nación*, which in 2003 had an average daily circulation of 200,000 copies nationwide, carried almost daily coverage of both educational and environmental issues. The paper also published occasional environmental education supplements (called *Aula Verde* or 'Green Classroom') and weekly science education supplements (called *Zurquí*) which were designed for use as study guides at home or in classrooms. Many educators and policy makers working at the national level noted that these diverse means of communicating environmental ideas were increasingly important in building public support for conservation and sustainable development initiatives.

Regional and International Influences

In addition to the variety of perspectives and practices of environmental education found within state-funded programmes during the time of this research, an extensive network of NGOs and other organisations employed a similarly diverse set of definitions of environmental education, and targeted a wide range of populations. As noted previously, many of these organisations have their roots in long-standing relationships with US scientists and scientific organisations. While many groups worked exclusively with school children, others directed their programming at farmers, communities sited near conservation areas, or tourists. In particular, influential international scientific research institutions and conservation organisations – each with their own conceptualisations of environmental education – seemed to have particularly strong impacts on the content and styles of implementation of programmes.

External actors have also played a significant role in the national economy, education provision, and changes to the structure of the state education bureaucracy more broadly. Indeed, Costa Rican leaders have a long history of turning to international bodies and fellow nations – most notably the United States – to negotiate trade relationships, aid, and other forms of exchange (cf. Honey 1994). Since the 1950s, the state has also borrowed heavily from international lenders in order to finance the development of its agro-export economy and domestic infrastructure and to mediate the impacts of unstable world markets for coffee and bananas. In 1963, Costa Rica solidified relations with other nations in the region by joining the Central American Common Market. This expanded its duty-free export zone to encompass the entire region and offered new opportunities for export-led development. Membership in the market brought explosive growth throughout the 1960s and 1970s, but this was brought to a standstill by the national economic crisis between 1979 and 1982 (Biesanz et al. 1999: 61).

The crisis was most likely precipitated by a number of factors, including increasing prices for imported petroleum, a sharp decline in trade within the Central American Common Market due to the civil wars in Nicaragua and El Salvador, reduced demand for Costa Rican exports, domestic inflation, and high interest rates on existing foreign debt (Biesanz et al. 1999: 49). International lenders such as the World Bank, however, blamed the crisis largely on the state's anti-export bias (historically, the state had emphasised the development of domestic or regional markets over broader international trade) and uncontrolled growth of the public sector (Carnoy and Torres 1994: 68). Lenders therefore refused to provide further loans unless the state agreed to streamline the existing bureaucracy, encourage development of private enterprise, and become more open to global trade. The Costa Rican government was forced to significantly decrease public spending – especially on social services – and to encourage 'more freedom for market forces and a leaner state whose public services should be run more according to management principles of private business and industry' (Lauglo 1996: 222). Between 1985 and 1994, successive governments signed three structural adjustment agreements which had far-reaching impacts on state provision of many kinds of social programmes, and especially on education.

The World Bank's emphasis on privatisation of public services, in particular, made the Costa Rican state's traditionally heavy spending on education an almost immediate focus of structural adjustment requirements. A policy of requiring cuts to state spending on education may seem contradictory in this case, particularly given that aid organisations generally agree that increasing spending on education is an effective means of promoting economic and social development. Groups such as the Economic Commission of Latin America (ECLA), for example, have actively sought to increase education investment throughout the region in order to stimulate 'the labor force productivity needed to increase international competitiveness and to secure a sustained economic dynamism' (ECLA, 1989: 278 quoted in Morales-Gómez and Torres 1992: 1). As Booth notes, however: 'there is evidence that the World Bank and other lenders considered Costa Rica's educational development as *excessive* for a Third World nation and that it might have to "give up some of what it had achieved"' (1998: 95, original emphasis).

The impacts of structural adjustment agreements remain visible today, as does the massive foreign debt accrued by the government during this period:

> 'Between 1983 and 1989, Costa Rica received almost US $2 billion in financial and technical assistance funds from bilateral and multilateral sources.... Such massive foreign assistance has made Costa Rica more dependent than ever on foreign advice in defining how the economy and society will develop in the future. Since Costa Rica's governments are now convinced that economic growth – hence their legitimacy – depends on foreign aid, and foreign aid agencies, in turn, require certain conditions to be met, Costa Rica is gradually turning into these agencies' vision of its economy and society.... this increased reliance on foreign aid and foreign expertise is also shaping the education system.' (Carnoy and Torres 1994: 74)

There is a considerable body of evidence which suggests that organisations like the World Bank have played a powerful role in changes to Costa Rica's national education system and other social services. It would be a gross oversimplification to suggest that the nation's educational policy has been wholly determined by international aid organisations, however, because structural adjustment programmes have also been a source of intense domestic debate. These conflicts have served to highlight tensions between domestic actors who favour increasing privatisation and believe that the state bureaucracy's massive size is an obstacle to development, and those who mourn the state's decreasing influence and maintain that state involvement in social welfare issues is a prerequisite for national development (Biesanz et al. 1999: 61; see also Nygren 1998).

In any case, the current state of the national education system can not be entirely blamed on structural adjustment. A number of the problems identified in the national education system can also be linked to other, mostly internal, factors. The state had already begun to decrease spending on education between 1980 and 1982, for example, in response to an earlier recession period and prior to the introduction of structural adjustment (Carnoy and Torres 1994: 80). Whatever the relative effects of internal and external pressures, it is clear that by the 1980s the state education system was experiencing severe financial and organisational difficulties. An internal assessment conducted by the Ministry of Education in 1988, for instance, revealed on-going problems such as the diminishing education budget, insufficient planning and co-ordination, deteriorating physical infrastructure (especially in schools) and insufficient teaching personnel, as well as a serious need to update curricula (Fischel Volio 1992).

Throughout its economic struggles, however, the Costa Rican state has also received significant support for education through a variety of other avenues and its extensive organisational and international links. The nation joined UNESCO in 1950, for example, and is the organisation's regional base in Central America.

Since the 1980s, however, international support for education in Costa Rica has increasingly centred around education about environmental issues, rather than on issues such as literacy and general education provision that are of much greater concern elsewhere in the region. Simultaneous with growing international investment intended to minimise the environmental impacts of agro-export production of bananas, coffee and beef throughout Latin America (cf. Price 1994), international organisations also began to offer support for national environmental educational efforts in the early 1980s. The US Peace Corps, for instance, assisted in the creation and publication of one of the Ministry of Education's first teaching guides for environmental education in 1991, in addition to offering support and technical expertise for the establishment of the national park system.

Indeed, the promotion of environmental education and its links to conservation can be understood as a strategically successful move by the state to attract international interest and funding for the educational sector more widely. Whereas international interest in the rest of the region has historically centred around social and economic concerns such as poverty, human rights, and ethnic conflict, Costa Rica's history of peace, stable democratic governance and relatively high standard of living have made the nation of marginal interest to many international groups. Organisations such as Action Aid, CARE and Oxfam, for example, do not operate in the country because it is not considered to be as desperately in need of development aid as other nations.[15] Even USAID, which has been deeply involved in aid funding to Costa Rica since the mid-1960s (Morgan 1993: 51), closed its Costa Rican office in 1996 when the organisation judged that it had achieved 'advanced developing country' status (US Department of State 1996). Active international interest in conservation and education, in contrast, continues to attract resources not only for land purchase, forest protection, research and tourism, but also – both directly and indirectly – for education.

NGOs – be they foreign or domestic, large or small – are also demonstrably powerful in political, economic and social terms in Costa Rica, and they have taken an active role in the formation of national environment and development policy, often by supporting pro-conservation government actions and strongly criticising neo-liberal economic and political trends in the country (O'Brien 1997: 185–189). This wide range of organisations and efforts are thus deeply embedded in processes of conservation, research, development, and environmental education provision in Costa Rica. Educational efforts by the state, and national and international NGOs in turn combine – directly and indirectly – with the actions of other privately-funded actors, including private schools, ecotourism businesses and individual citizens working as environmentalists, scientists, educators and campaigners. Equally, as sustainable development initiatives have received increasing attention at the international level and in Costa Rica, the ideological, political and economic connections

---

[15] The war in Iraq has also had implications for the availability of funding. In 2004, DFID closed all but one (Nicaragua) of its country programmes in Latin America in order to re-channel funds to the Middle East.

between forest conservation, education and community development have continued to strengthen.

Attention to these links internationally has also resulted in higher levels of available funding for sustainable development projects, and has encouraged the re-orientation of projects initiated by both the Costa Rican state and domestic and international NGOs. In particular, many conservation organisations in Costa Rica increasingly see their role as encompassing not only forest protection and research, but also public education. As so much of the social and economic support for education in Costa Rica has centred around environmental issues, environmental education has become an important point of intersection between state, non-state, domestic and international actors as well as a focal point for debates about the content and aims of education and its relationships to sustainable development.

References

Augelli, J. P. (1987). Costa Rica's frontier legacy. *Geographical Review, 77*(1), 1–16.
Barraza, L., Duque-Aristizábal, A. M., & Rebolledo, G. (2003). Environmental education: From policy to practice. *Environmental Education Research, 9*(3), 347–357.
Benavides Jiménez, C. R. (2009). *Informe final de gestion: Mayo 2006 - Setiembre 2009*. Costa Rica: Instituto Costarricense de Turismo.
Biesanz, M. H., Biesanz, R., & Biesanz, K. Z. (1999). *The Ticos: Culture and social change in Costa Rica*. Boulder: Lynne Rienner Publishers.
Blum, N. (2008). Environmental education in Costa Rica: Building a framework for sustainable development? *International Journal of Educational Development, 28*(3), 348–358.
Booth, J. A. (1998). *Costa Rica: Quest for democracy*. Boulder: Westview Press.
Boza, M. A. (1993). Conservation in action: Past, present and future of the National Park System of Costa Rica. *Conservation Biology, 7*(2), 239–247.
Brockett, C. D., & Gottfried, R. R. (2002). State policies and the preservation of forest cover: Lesson from contrasting public-policy regimes in Costa Rica. *Latin American Research Review, 37*(1), 7–40.
Campbell, L. (2002). Conservation narratives in Costa Rica: Conflict and co-existence. *Development and Change, 33*, 29–56.
Carnoy, M., & Torres, C. A. (1994). Educational change and structural adjustment: A case study of Costa Rica. In J. Samoff (Ed.), *Coping with crisis: Austerity, adjustment and human resources* (pp. 64–99). London: Cassell and UNESCO.
Carriere, J. (1991). The crisis in Costa Rica: An ecological perspective. In M. Redclift & D. Goodman (Eds.), *Environment and development in Latin America: The politics of sustainability* (pp. 184–204). Manchester: Manchester University Press.
Castillo, M.S. (2006). *Los valores en el planamiento didáctico: Eje transversal del curriculo Costarricense*. Despacho del Ministro – Programa Nacional de Formación en Valores. San José, Costa Rica: Ministerio de Educación Publica.
Coughlin, M. D., Jr. (1993). Using the Merck-INBio agreement to clarify the convention on biological diversity. *Columbia Journal of Transnational Law, 31*(2), 337–375.
Dutschke, M., & Michaelowa, A. (2000). Climate cooperation as development policy: The case of Costa Rica. *International Journal of Sustainable Development, 3*(1), 63–94.
Eakin, M. C. (1999). The origins of modern science in Costa Rica: The Instituto Físico-Geográfico Nacional, 1887–1904. *Latin American Research Review, 34*(1), 123–150.
Edelman, M. (1995). Rethinking the hamburger thesis: Deforestation and the crisis of Central America's beef exports. In M. Painter & W. H. Durham (Eds.), *The social causes of*

References

environmental destruction in Latin America (pp. 25–62). Ann Arbor: University of Michigan Press.

Edelman, M., & Kenen, J. (Eds.). (1989). *The Costa Rica reader*. New York: Grove Weidenfeld.

Evans, S. (1999). *The green republic: A conservation history of Costa Rica*. Austin: University of Texas Press.

Fischel Volio, A. (1987). *Consenso y represión: Una interpretación socio-política de la educación Costarricense*. San José: Editorial Costa Rica.

Fischel Volio, A. (1992). Costa Rica: Education and politics – a historical perspective. In D. A. Morales-Gómez & C. A. Torres (Eds.), *Education, policy and social change: Experiences from Latin America* (pp. 141–156). London: Praeger.

Fournier, L. A. (1991). *Desarrollo y perspectivas del movimiento conservacionista Costarricense*. San José: Editorial de la Universidad de Costa Rica.

Garcia Gómez, J. (2000). Modelo, realidad y posibilidades de la transversalidad: El caso de Valencia. *Tópicos en Educación Ambiental, 2*(6), 53–62.

González Gaudiano, E. (2000). Los desafíos de la transversalidad en el currículum de la educación básica en México. *Tópicos en Educación Ambiental, 2*(6), 63–69.

Heyneman, S. P. (2003). The history and problems in the making of education policy at the World Bank, 1960–2000. *International Journal of Educational Development, 3*(1), 315–337.

Holdridge, L. (1947). Determination of world plant formations from simple climatic data. *Science, 105*(2727), 367–368.

Holdridge, L. (1967). *Life zone ecology*. San José: Tropical Science Center.

Honey, M. (1994). *Hostile acts: US policy in Costa Rica in the 1980s*. Gainesville: University Press of Florida.

Honey, M. (1999). *Ecotourism and sustainable development: Who owns paradise?* Washington, DC: Island Press.

Huckle, J., & Sterling, S. (Eds.). (1996). *Education for sustainability*. London: Earthscan.

ICT. (2003). *Anuario estadístico del turismo* (Government Report). San José, Costa Rica: Instituto Costarricense de Turismo.

INEC [Instituto Nacional de Estadística y Censos]. (2004). *Cálculo de población: por provincia, cantón y distrito al 1° de Enero de 2004* (Government Report). San José, Costa Rica: INEC.

Janzen, D. (Ed.). (1983). *Costa Rican natural history*. Chicago: University of Chicago Press.

La Nación. (2003a). *Sin control tráfico ilegal de pieles de caimán* (pp. 4A–5A). San José: La Nación.

La Nación. (2003b). *Costa Rica descuida su imagen ecológica: MINAE reconoce que necesita más dinero y personal* (p. 1A). San José: La Nación.

La Nación. (2003c). *Dimitió Astrid Fischel* (pp. 1A–5A). San José: La Nación.

Lauglo, J. (1996). Banking on education and the uses of research: A critique of World Bank priorities and strategies for education. *International Journal of Educational Development, 16*(3), 221–233.

Lavery, A., & Smyth, J. C. (2003). Developing environmental education, a review of a Scottish project: International and political influences. *Environmental Education Research, 9*(3), 359–383.

Lencastre, M. P. (2000). Transversalización curricular y sustentabilidad: Contribución para la teoría y práctica de la formación de maestros. *Tópicos en Educación Ambiental, 2*(6), 7–18.

Luzzi, D. (2000). La educación ambiental formal en la educación general básica en Argentina. *Tópicos en Educación Ambiental, 2*(6), 35–52.

Martin, P. (1996). A WWF view of education and the role of NGOs. In J. Huckle & S. Sterling (Eds.), *Education for sustainability* (pp. 40–51). London: Earthscan Publications.

Martinez, M. (1998). *Consideraciones teóricas sobre la educación en valores*. Government policy paper. San José, Costa Rica: MEP SIMED.

Mayorga, G., Sánchez, J., & Palmer, P. (2004). Taking care of Sibü's gifts. In S. Palmer & I. Molina (Eds.), *The Costa Rica reader: History, culture and politics* (pp. 264–275). Durham: Duke University Press.

McKeown, R., & Hopkins, C. (2003). EE≠ESD: Defusing the worry. *Environmental Education Research, 9*(1), 117–128.

MEP and OEA. (2002a). *'Desarrollo en armonía con la naturaleza'. Módulo No. 1: Módulo de educación ambiental para docentes de I y II ciclos de la educación básica*. Curriculum document. San José, Costa Rica: Ministerio de Educación Publica.

MEP and OEA. (2002b). *'Intervención humana en el entorno'. Módulo No. 2: Módulo de educación ambiental para docentes de I y II ciclos de la educación básica*. Curriculum document. San José, Costa Rica: Ministerio de Educación Publica.

MEP and OEA. (2002c). *'Mediación pedagógica de la educación ambiental'. Módulo No. 3: Módulo de educación ambiental para docentes de I y II ciclos de la educación básica*. Curriculum document. San José, Costa Rica: Ministerio de Educación Publica.

MEP. (2004). *Programa de formación en valores*. Government policy document. San José, Costa Rica: Despacho del señor Ministro de Educación – PROMECE.

MEP [Ministerio de Educación Publica]. (2002). *Los temas transversales en el trabajo de aula*. Curriculum document. San José, Costa Rica: Ministerio de Educación Publica.

MINAE [Ministerio de Ambiente y Energía]. (2000). *Estrategia de conservación y uso sostenible de la biodiversidad, Área de Conservación Arenal-Tilarán*. Government strategy paper. San José, Costa Rica: Ministerio de Ambiente y Energía.

Mirón, L. (1989). Costa Rican education: Making democracy work. In M. Edelman & J. Kenen (Eds.), *The Costa Rica reader*. New York: Grove Weidenfeld.

Morales-Gómez, D., & Torres, C. A. (Eds.). (1992). *Education, policy and social change: Experiences from Latin America*. London: Praeger.

Morgan, L. (1993). *Community participation in health: The politics of primary care in Costa Rica*. Cambridge: Cambridge University Press.

Nygren, A. (1998). Environment as discourse: Searching for sustainable development in Costa Rica. *Environmental Values, 7*, 201–222.

O'Brien, P. J. (1997). Global processes and the politics of sustainable development in Colombia and Costa Rica. In R. M. Auty & K. Brown (Eds.), *Approaches to sustainable development* (pp. 169–194). London: Pinter.

OEA [Oficina de Educación Ambiental]. (2002). Marco conceptual, legal y áreas de acción de la Oficina de Educación Ambiental. Unpublished internal document. San José, Costa Rica: Oficina de Educación Ambiental/Ministerio de Educación Publica.

Pérez, H. (1981). *Educación y desarrollo: Reto a la sociedad Costarricense*. San José: Editorial Costa Rica.

Price, M. (1994). Ecopolitics and environmental nongovernmental organizations in Latin America. *Geographical Review, 84*(1), 42–58.

Proyecto Estado de la Nación. (2002). *Estado de la nación en desarrollo humano sostenible: Octavo informe*. San José: Proyecto Estado de la Nación.

Proyecto Estado de la Nación. (2004). *Estado de la nación en desarrollo humano sostenible: Décimo informe*. San José: Proyecto Estado de la Nación.

Reding, A. (1986). Voices from Costa Rica. *World Policy Journal, 3*(2), 317–345.

Reigota, M. (2000). La transversalidad en Brasil: Una banalización neoconservadora de una propuesta pedagógica radical. *Tópicos en Educación Ambiental, 2*(6), 19–26.

Roth, E. (2000). Medio ambiente como transversal en la educación formal: Algunos apuntes en la experiencia Boliviana. *Tópicos en Educación Ambiental, 2*(6), 27–34.

SINAC-MINAE. (2003). Informe nacional sobre el Sistema de Áreas Silvestres Protegidas. Unpublished report. San José, Costa Rica: Gerencia de Áreas Silvestres Protegidas, Sistema Nacional de Áreas de Conservación (SINAC), Ministerio del Ambiente y Energía.

Sterling, S. (2001). *Sustainable education: Re-visioning learning and change* (Schumacher Briefing Number 6). Totnes: Green Books Ltd.

Tico Times. (2003). *'Minister: Teachers getting their pay'* (p. 6). San José: The Tico Times.

Twombly, S. (1997). Curricular reform and the changing social role of public higher education in Costa Rica. *Higher Education, 33*, 1–28.

UNDP. (2005). *International cooperation at a crossroads: Aid, trade and security in an unequal world* (Human Development Report 2005). New York: Oxford University Press and United Nations Development Programme.

References

UNESCO and RNTC [Radio Nederland Training Centre]. (2000). *Hacia una cultura de prevención de desastres: Guía para docentes*. Curriculum document. San José, Costa Rica: UNESCO and Radio Nederland Training Centre.

UNIMER. (2002). *Estudio nacional sobre valores ambientales de las y los Costarricenses*. San José: UNIMER International, La Nación, Proctor & Gamble, Amanco and the Universidad Latinoamericana de Ciencia y Tecnología.

US Department of State. (1996). *Background notes: Costa Rica*. Washington, DC: Bureau of Inter-American Affairs.

Vargas, J., Mayorga, R., Leiva, C., Morales, A., Moyorga, G., Sánchez, J., Palacios, E., & Morales, C. (1999). Costa Rica. In D.A. Posey (Ed.), *Cultural and spiritual values of biodiversity* (pp. 135–136). London: UNEP and Intermediate Technology Productions.

Vaughn, C., & Rodriguez, C. M. (1997). Managing beyond borders: The Costa Rican National System of Conservation Areas (SINAC). In G. K. Meffe & C. R. Carroll (Eds.), *Principles of conservation biology* (pp. 441–451). Sunderland: Sinauer Associates.

Wallace, D. (1992). *The Quetzal and the Macaw: The story of Costa Rica's national parks*. San Francisco: Sierra Club Books.

Chapter 3
Environmental Education in Schools[*]

Abstract Research has given critical attention to diverse theories and practices of environmental education, but has tended to take a narrow focus on specific curricula and policies or on activities within strictly defined sites such as schools, classrooms or protected areas. In contrast, this research argues that greater attention needs to be given to the broader social, economic and political contexts in which these initiatives take place, as well as to how they impact upon educational practice. The chapter begins with an introduction to the community of Monteverde, Costa Rica, where the research was conducted. The discussion then examines how the content and goals of environmental education programme in local schools are strongly linked to a range of wider social and economic relationships, and explores the ways in which these impact upon educational practice. In particular, despite their diverse sizes and relationships to the state, local schools faced many of the same difficulties in promoting environmental learning, including meeting the expectations of the state, parents, and employers; structural concerns such as high teacher turnover and limited resources; and the demands of the state's heavily content-based curriculum and strict national examination requirements.

Keywords Environmental education • Community • Schools • Science education • Transformative learning

The existing literature from environmental education and related fields suggests that formal educational institutions, such as state and privately funded schools, are important sites for environmental learning.[1] Indeed, much of the literature and

---

[*] A version of this chapter was previously published as Blum (2008).
[1] As I outlined in Chap. 1, the discussion throughout this book refers largely to the English-language literature related to these issues. Much more research is needed to explore how these issues play out in other contexts around the world.

policy related to environmental education has focused exclusively on learning in these formal settings. However, it is also important to recognise that schools do not operate in isolation from the communities and nations in which they are located. For instance, they may have quite different relationships to state education systems, as well as to diverse local perspectives, interests and needs. As a result, it is difficult, and probably not even desirable, to research environmental learning in the formal sector without also trying to understand these wider relationships and influences, and their impacts.

Furthermore, in much of the literature on education and development, as well as in many international and national development policies and agendas, there is an assumed connection between the provision of education and increasing levels of development. This connection is rooted in a long-standing tradition of international research and policy on education and development (cf. Schultz 1961; Sen 1999; Nussbaum 2000), as well as in the work of international aid organisations and funders (e.g. through global initiatives such as Education for All; see Little et al. 1994; UNESCO 2000). While debates continue about the exact linkages and the kinds of development that they might encourage (e.g. economic, social or sustainable development), the central idea remains largely unchanged: more education = increased development.

Teachers are therefore commonly viewed by policy makers and general publics as key actors in the provision of the knowledge and skills that future generations will need to achieve sustainable development. In other words, they are entrusted with the implementation of educational (development) policy 'on the ground'. This assumption, however, rarely takes into account that they must work within the boundaries and requirements of a state's educational policy, and also within local economic and social norms and conditions. Teachers' negotiations of educational policy and practice therefore depend to a large extent on the make-up of individual schools (e.g. public or private, large or small) and how these are located within complicated national and local economic and social contexts. As this chapter will explore, the diverse ways in which schools are located within these complex relationships in Monteverde highlights the profound impacts that these can have on both perspectives on, and the implementation of, environmental education programmes.

Welcome to Monteverde

The Monteverde region is located in north-western Costa Rica within the Tilarán mountain range, straddling the Continental Divide, and roughly 150 km from the nation's capital city, San José. It is one of Costa Rica's most popular tourist destinations, as well as perhaps its most highly researched. In the 1990s, it was designated by the Costa Rican Tourism Institute (ICT) as one of four sites for special focus by the growing national ecotourism industry, and by 2003 the region was attracting an estimated 200,000 visitors each year (MVI 2002), including researchers, conservationists, bird enthusiasts, international students and adventure tourists. Tourism is also the most important local employer either directly – through jobs in hotels, tour or transport companies or at local attractions – or indirectly – for those who are employed by other local services. By 1992, the annual economic impacts of the local tourism industry were estimated at US$5 million for a regional population of approximately 4,000 inhabitants (Burlingame 2000: 376), and the industry has continued to grow rapidly since then. The region is also known internationally as a site for innovation. It was the first place in Costa Rica, for example, to participate in a 'debt-for-nature' swap, and was also the site of the establishment of the nation's first conservation easement.[2] As a community, Monteverde is as emblematic of a 'green' destination as the nation of Costa Rica itself, at least partly due to the heavy media promotion of the community as a site for ecotourism.

The rare cloud forests of the region have proven central to the economic wellbeing of the community by attracting increasing numbers of visitors since the 1970s. The expansive growth of the local tourism industry has had significant impacts on the social and economic life of the community. While some residents have welcomed the growth of the tourism industry and its resulting affluence, others believe that it has brought serious problems with it, including alcohol and drug abuse and the more general erosion of social values. Nevertheless, Monteverde still relies on its reputation for conservation and biodiversity in order to attract tourism revenues, and remains largely rural in character. Visitors enjoy the 'small town' feel of the region and often visit as a break from the mass tourism of beach communities. Those popular tourism destinations, in contrast, are often characterised by the unregulated development of hotels and restaurants, as well as a general lack of water and waste management which have resulted in high levels of contamination along the coasts.

Monteverde's forests have also been the greatest attraction for scientists, conservationists, artists and businesspeople – many of whom have come to live and work in the region permanently. Increasing numbers of annual visitors have put particular pressure on protected areas and local tourism businesses to provide services. On local protected area – the Monteverde Reserve – for example, received only 471 visitors in 1973–1974, but since the 1990s has received an average of 50,000 tourists per year (Aylward et al. 1996: 325–327). Similarly, the smaller Santa Elena Reserve has experienced a significant increase in visitation, jumping from 3,100 visitors during its first year of operation to more than 13,000 per year in the mid-1990s (Burlingame 2000: 368). By 2002, the estimated number of visitors to the region as a whole had risen to approximately 200,000 per year (MVI 2002). Regionally, the period of greatest growth in both the tourism industry and in terms of local population has occurred since 1986. In contrast to the rest of the country – which at the time of this research was experiencing an estimated 2.8% rate of annual

---

[2] For more on the debt-for-nature swap system, see Burlingame (2000: 362-363) and Dutschke and Michaelowa (2000). Conservation easements are legal arrangements, first employed in the US, which establish permanent restrictions on the possible uses of a property and its resources. The concept was pioneered in Latin America by CEDARENA, a legal NGO based in San José, Costa Rica. See http://www.cedarena.org/landtrust.

population growth – the Monteverde region showed a 7% rate of growth, much of it from in-migration either from other parts of the nation or from other countries (MVI 2002). In a 2002 survey conducted by the Monteverde Institute, a local research organisation, 56% of the residents interviewed categorized themselves as originating from outside the Monteverde region, and 44% from within it. When disaggregated by settlement, the data shows that this growth has affected local settlements in diverse ways, with some villages showing a very high concentration of residents who were born outside the region (e.g. 81% in the village of Monteverde) and others much lower (e.g. an estimated 23% in nearby La Lindora). Overall, the highest rates of population growth in the region have occurred in the commercial centres of Santa Elena (growing from 360 residents in 1986 to 1,866 in 2002) and Cerro Plano (from 186 residents in 1986 to 1,266 in 2002) (MVI 2002).

The resulting mixture of nationalities, languages and perspectives represented within the local population, in turn, has had interesting implications for social relationships in the community as a whole and for environmental education in particular. There are, for example, observable divisions between the social lives of Costa Rican residents (which tend to focus on family relationships) and their foreign counterparts (centred on religious affiliations or, in the absence of extended families, are constructed around work associations or shared nationality/language), and both positive impressions and negative stereotypes can be found on both sides of this division.

Given the diversity of the region's population, language has proven especially important in the local context. Many community members, especially long-term settlers, are fluent in more than one language – most commonly Spanish and English. On both an individual and an organisational level, this fluency can help to form the links which makes collaboration possible, but a lack of it can also result in conflict or misunderstanding. Not just individuals, but also locally important nongovernmental organisations (NGOs) tend to be categorised by residents as either '*gringo*' or '*Tico*' – differences which have had significant impacts on the potential for both collaboration and conflict.[3] As a result, several local institutions have struggled to engage the participation of more than one linguistic group. So-called 'gringo' organisations, for example, were often originally established by US immigrants, and in 2003 a number of Spanish-speaking residents claimed that these organisations continued to be dominated by 'gringo' ways of thinking about and responding to local issues.

At the same time, the vast majority of community members commonly expressed a shared commitment to consensus-building and peaceful collaboration. Costa Rican residents tended to link this commitment to national cultural and social norms, including the desire to *quedar bien* (as discussed in the previous chapter). North American settlers – many of whom were drawn to settle in Costa Rica precisely because of its reputation for peace – also strongly supported consensus-building and

---

[3]*Tico* is the affectionate name which Costa Ricans use to identify themselves. It refers to the idiosyncratic use in Costa Rican Spanish of the diminutive *–tico*, rather than the more common Spanish diminutive *–ito*.

the peaceful search for solutions to local concerns. This was particularly true of the local Quaker community, whose members – originally from the United States – linked their commitment to peace with religious obligations. Local debates I witnessed during the course of this research were conducted with strongly diplomatic and collaborative vocabulary, and local decision-making and organisational culture centred on participation in committees, seminars and workshops. Personal or organisational conflicts, on the other hand, tended to remain hidden and were often only expressed privately.

The relatively recent settlement of both Costa Ricans and foreigners in the region – with the first few Costa Rican families arriving as late as the 1930s, the first North Americans in the early 1950s, and continuing in-migration in the present day – also tended to blur the boundary between 'insiders' and 'outsiders', and community members expressed relatively little opposition to the arrival of new settlers or ideas. It has also meant that arguments for the protection of local forests centre around the community's development as a premier site for international scientific research, successful forest protection, and ecotourism. This is in sharp contrast to discourses of long-standing or indigenous connection to land which are so often at the centre of debates about local resource management elsewhere in the world. Those who were born in the region as well as those who arrived because of its 'green' reputation expressed real pride at the community's history of preservation, and talked openly of a desire to conserve a local way of life which is embedded in, and reliant upon, maintaining the ecological balance of the surrounding forests. This balance is important both socially and economically, as without the forests to attract tourism and scientific interest, there is little else that could support the local population at its current size.

Although it is by no means a typical Costa Rican community, with its diverse population, relative affluence and successful history of conservation, these characteristics serve to make the links between conservation, tourism and education very clear in Monteverde. Resident researchers, conservationists, educators, business interests and a wide variety of local organisations take an active interest in the maintenance of the area's ecological health. Perhaps most importantly for this book, education about these local resources is also very actively promoted within the community. As is the case throughout the country, state schools in the region administer environmental education programming as part of the approved national curriculum. In addition, in Monteverde during the time of this research there were four influential local NGOs and two private schools, as well as a large number of smaller informal groups, committees and commissions, working in conservation and environmental education. While these groups did not always agree on the best ways to deal with local environment and development issues, their sometimes conflicting perspectives and practices formed an important network of discussion, effort, and innovation which may be the key to the community's success.

Before moving on, it is necessary to clarify an important issue of identification. The name 'Monteverde' is commonly used in the local context to refer to a number of different entities. Firstly, there is the legal political entity, District 10, of the *canton* of Puntarenas, which itself lies within the province of Puntarenas. The name

'Monteverde' is also used to describe a small settlement originally established by a group of US Quakers in 1951. Local residents also routinely use the name to refer to what can be perhaps more clearly labelled the 'Monteverde zone' or 'Monteverde region'. This larger geographical area is commonly understood by local residents and researchers to encompass the three relatively large settlements of Monteverde, Cerro Plano and Santa Elena (sites for the most intensive tourism and conservation efforts in the region), as well as the three main protected areas and smaller settlements on both sides of the Continental Divide. In its broadest usage, local residents speak simply of 'the community' (*la comunidad*) when referring to local events, happenings or sites, and the social and economic relationships which bind geographically separate settlements. I have followed this local pattern of usage throughout the book, so the terms 'Monteverde' and 'community' should be understood to geographically encompass the villages of Santa Elena, Cerro Plano and Monteverde while also acknowledging the strong social and economic ties with smaller settlements nearby. When referring to the Quaker settlement specifically, I have referred to it as the 'village of Monteverde'.

Formal Education in Monteverde

Formal state schooling has been available in the Monteverde region on a limited basis since at least the early twentieth century. The population at this time was largely composed of geographically-isolated family farms which practised subsistence agriculture and small-scale local trade. Nevertheless, by the time the Quaker settlers arrived from the US in 1951, the villages of Santa Elena and Cerro Plano each already had a school providing instruction for the first two grade levels (see Monteverde Friends Meeting 2001). The Quaker settlers quickly established their own school, the Monteverde Friends' School, in order to provide lessons for the community's children. As the region's population steadily grew, the two state primary schools expanded their grade level offerings, and in 1977 the state secondary school (*Colegio*) was opened in Santa Elena. In addition to the Friends' school, the Seventh Day Adventist Church also later established a private school, and in 1991 a group of American residents founded the Cloudforest School, a private 'environmental education' school.

During the time of this research, the teaching staff employed by each of these schools varied greatly. The majority of state school teachers, for example, were Spanish-speaking Costa Rican nationals, although relatively few were originally from the Monteverde region. Newly qualified teachers in the state system are allocated teaching positions by the Ministry of Education on the basis of their relative achievement on professional qualifying examinations. Those with the highest exam scores are given first choice of job openings in state schools, and most frequently choose to take positions at urban schools in the Central Valley. State teachers in Monteverde attributed this tendency to the better access to resources and teaching materials available to educators in these schools, and also to the fact that many teachers (and the majority of university graduates) are from the Central Valley themselves and prefer to live and work near family and friends. As a result, teachers arriving to take up work in rural areas like Monteverde may plan to stay in the region only until they can be re-assigned to a school in a more favourable location, and they often have relatively little knowledge of the local community or environment when they arrive. Private schools in Monteverde, on the other hand, largely recruited and employed teachers from the United States because teaching was conducted largely in English, and because US teachers were more likely to have received training in the concept-based and 'child centred' teaching approaches favoured by those schools. Similarly to their state school counterparts, however, these teachers often arrive with little knowledge of the local environment and may only plan to stay for 1 or 2 years.

For their part, local parents make choices about enrolling their children in one of the local schools for a variety of reasons, both explicit and implicit. Attendance at local English-language private schools, for example, is highly competitive among parents whose first language is Spanish because of the common belief that English fluency is an asset for future employment in the local tourism industry. Parents who are actively involved in local environmental issues – especially through ownership of local tourism attractions or work with conservation organisations – commonly choose to send their children to the Cloudforest School both for its environmental education orientation and because classes are taught in English. Other parents choose to place their children at the Monteverde Friends' School in order to provide other kinds of learning opportunities, especially related to its religious orientation, as well as for English language learning. Access to such privately-funded education is largely contingent on a family's resources, but the schools do provide need-based grants to a substantial number of students.

Local parents also make strategic decisions about school enrolment in response to pressures from the national system of assessment. Students in both state and privately-funded schools throughout Costa Rica are required to take national exams at the end of the sixth, ninth, and eleventh grade years. These exams are incredibly important in the educational life of each student. Admission to secondary school is contingent on passage of the sixth grade exams, for example, and the eleventh grade exams determine whether or not students receive a secondary school qualification and thus also determine a student's eligibility and relative opportunity to go into higher education. For this reason, parents often send their children to one of the local private schools for a few years in order to learn English, before moving them to a state school in time for preparation for the national exams (administered in Spanish).

Environmental Education in Monteverde's Schools

In 2003, environmental education was provided in almost all local schools, although to varying degrees and in differing ways. While the diversity of school types in the region – state, private, religious, small, large – suggests that each one has a distinct set of concerns regarding student progress and success, they also faced many of the same difficulties in terms of educational practice. These included the negotiation of definitions of environmental education and ways of implementing it, as well as the need to meet the expectations of the state, parents, and employers in the local tourism industry. Each school also struggled with structural concerns such as high teacher turnover and limited class time, resources and teacher training. In addition, teachers in both state and private schools identified the state's heavily content-based curriculum and strict national examination requirements as particular problems. These last two issues had especially strong impacts on environmental education because – although it is placed in a prominent position by the state and sometimes by the schools themselves – in practice, environmental learning was often marginalized in favour of meeting the assessment requirements of the Ministry of Education. In each case, the design and implementation of environmental education programmes was mediated by each school's unique position in relationship to both the state education bureaucracy and to a variety of community interests.

State Primary Schools and *Temas Transversales*

There were two state primary schools in Monteverde at the time of this research: *Escuela Santa Elena* had the largest student population (approximately 300 students) and newest facility, while *Escuela Cerro Plano* was a much smaller site (only 3 classrooms and 97 enrolled students). In both cases, the schools followed the national curriculum and structured the school day in accordance with the common national practice of providing two half-day sessions. Environmental education in both schools was provided in the formal curriculum through the Ministry of Education's 'transversal themes' (*temas transversales*) and less formally by the environmental education co-ordinator from the Monteverde Reserve (whose work is discussed in detail in Chap. 4). In 2003, there were four of these themes: 'human rights, democracy and peace', 'building a culture of environmentalism and sustainable development', 'health education', and 'sex education'. The second, 'building a culture of environmentalism and sustainable development', was intended to be used by teachers to 'integrate' environmental education themes into all areas of the curriculum.

The Ministry's publication on the themes describes in great detail the diverse range of required topics for each, as well as an extensive list of what it labels as important Costa Rican social values that should be used by teachers to promote the personal, social and ethical development of students. The list includes, but is not limited to: generosity, love, solidarity, responsibility, truth, non-violence, justice, liberty, citizen participation, tolerance, friendship, equality, peace, democracy, honesty, environmentalism, respect for others, and respect for diversity (MEP 2002: 25). State school teachers were instructed to use the themes and promote these values in a number of ways, such as by integrating them into studies of the basic subjects through inventive classroom activities, or by encouraging positive 'school cultures' in which teachers, administrators and students practice important social values in their interactions both inside and outside of the classroom (ibid: 12). The promotion of these positive values within schools was overwhelmingly supported by educators and parents, however many educators were also quick to point out that their practical implementation was often exceedingly difficult. Indeed, many educators working in state primary schools reported that the only environmental education actually happening in their classrooms was through the programmes provided by the Monteverde Reserve.

When asked about the problem of providing environmental education in their schools, local teachers identified a number of obstacles. Firstly, these included complaints about the general lack of resources as well as the heavy content requirements of the national curriculum. Many classrooms I visited during the fieldwork year, for example, were completely empty of learning materials such as textbooks, notebooks or posters. Even basic resources were often so scarce that teachers frequently paid for chalk and paper out of their own wages. Although parents are required to purchase Ministry-approved textbooks each year, they are often unable to afford them, so teachers compensated by making photocopies (again, out of their own pockets).

The physical infrastructure of state schools is also often in poor repair, at least partly as a result of successive declines to the state education budget since the 1980s. State schools across the country are usually simple one-story buildings of concrete blocks with tin roofs. Classrooms are similarly utilitarian and typically contain only a set of student desks and a chalkboard. When primary schools have their own libraries, these are often inadequate to the needs of the student body and composed of out-of-date texts. Declining state education budgets have also severely limited the construction of new schools to accommodate the nation's growing population. As a result, primary school teachers nationwide have had to 'double up' to provide classes for growing numbers of students in increasingly larger groups. This lack of schools (and of trained teachers to staff them) is at the root of the national practice of teaching two groups of primary students each day – a morning shift and afternoon shift – which extends the teaching day from early morning until early evening.[4] Teachers in Monteverde complained that this was utterly exhausting on a personal level, and also that it left little room for lesson planning, creative teaching, or professional development.

Educators in Monteverde's state primary schools also struggled with high rates of teacher turnover. During the 2003 academic year, for example, *Escuela Cerro Plano* had to dismiss two teachers – one of whom had also been acting as director for the previous 4 years – which left the school with a new director and only two teachers to give classes to nearly 100 students in several grade groups during the latter half of the academic year. This was in addition to the problems caused by a strike by the national teachers' unions earlier in the year, which had resulted in the loss of a month of teaching (see Chap. 2). It was in these rather austere circum-

---

[4]This is a common practice throughout Latin America and has significant consequences for environmental education in the region (see González Gaudiano 2007).

stances that state school teachers were expected to provide for the comprehensive study of Spanish, mathematics, social studies and general science required by the Ministry of Education. Perhaps due to such problems with supplies and training, or because of simple exhaustion, teachers in Monteverde (as elsewhere in the country) tended to rely on lectures and dictation to teach in the classroom, asking students to learn largely by rote memorisation. Student learning was then largely tested using only multiple choice or true-false tests – an approach to teaching and learning that stands in marked contrast to the creative, cross-curricular environmental learning promoted through the Ministry's transversal themes.

Training for Environmental Education

In 2003, the state provided only scarce training to teachers about environmental topics, largely due to its own limitations on available resources and staff. Similarly, the national Office of Environmental Education had little interaction with educators in Monteverde.

A small group of teachers at *Escuela Santa Elena* did, however, receive some training on the Ministry of Education's transversal themes during a half-day session at the school. The session was conducted by the school's own director and a representative from the Ministry's regional office in Puntarenas. The training day, in March 2003, was attended by about a dozen teachers representing several grade levels, as well as the environmental education co-ordinator from the Monteverde Reserve who kindly invited me to join them. After the session, these teachers would be responsible for passing on any information about the themes to their colleagues at the school. Classes were suspended for the day, and we met in a classroom on the school grounds. To begin the session, we were divided into small groups to discuss the material contained within the Ministry's publications on transversal themes. Afterwards, we came together again as a large group and outlined the central goals of the thematic teaching schema on a large piece of poster paper. These goals included teaching values, abilities, attitudes, and concepts. The director instructed her teachers: *Transversal themes are the theoretical framework through which learning objectives are achieved and values are taught. The themes do not need to be listed in your lesson plans, but they should be explicitly represented in the kinds of activities you do and topics that you teach.*

Much of the rest of the day's discussion similarly centred on values, and especially the perceived need for teachers to be involved in the lives of their students both inside and outside of the classroom, as well as in their wider communities. The director described how teachers should conduct 'research' when they arrived to work in a new community so that they could understand locally-appropriate values and standards of behaviour. *This has to be flexible depending on the community*, she added, *for example in Puntarenas drinking a beer is fine, but in many rural communities it would be considered completely unacceptable behaviour.* This message may have been especially appropriate for the group of teachers on this day, as a significant number of them were newly arrived in Monteverde themselves. The group as a whole agreed that developing this kind of local understanding is important not only for the sake of students, but also because they believed that their role as educators was to teach and guide students both inside and outside of school. They described themselves as responsible not just for 'giving out information' or 'knowledge' but, even more importantly, for facilitating the learning process and teaching values. In this way, the teachers strongly drew on wider national narratives about the role of education in the social and economic development both of individuals and of the nation.

In addition to this strong philosophical thread which ran through the discussion during the training session, there was also a good deal of practical discussion about useful ways to apply these ideas/requirements in the classroom. In one exercise, for example, we were divided into four small groups, each assigned a theme, and then asked to identify a learning objective from the national curriculum that could be used to elaborate it. My group was given the theme 'building a culture of environmentalism and sustainable development', so I was able to speak directly to the teachers in the group as we worked to find practical applications for the classroom.

Each group was instructed to use Bloom's (1984) taxonomy to plan activities that would move the students through the learning process from simple knowledge (*conocimiento*) of relevant information, to an understanding (*comprehensión*) of related issues, the application (*aplicación*) of this knowledge, and finally to an ability to critically analyse (*analizar*) the issues involved. After looking through the fourth grade curriculum, my group chose a learning objective that involved learning about biodiversity and its protection. For the first step (*conocimiento*) the students would be asked to classify a list of living things according to what they eat (e.g. omnivores, herbivores, and carnivores). Then, in the second step (*comprehensión*) they would make lists of the kinds of animals found in the students' own surroundings and relate this list back to the first step. This would be used as a way of beginning a class discussion of habitat destruction and species loss. The third step (*aplicación*) would build upon the knowledge gained so far by asking the students to create group projects in which they explored solutions to contemporary problems for animals in danger of extinction due to habitat loss or other environmental factors. The final step (*análisis*) would then require the students to conduct research on organisations, laws, and individuals already working on these problems, and to understand what they are doing to help.

Although all of the groups went about this detailed planning task with real willingness and energy, at several stages during the day the teachers also expressed serious misgivings about the complicated nature of the transversal theme requirements, and especially the additional time burden of planning for them. Some questioned whether and how the themes would be accounted for within national exams (especially given that there was no designated examination for them), and even whether learning of this kind could be effectively evaluated or measured at all. These queries were not so much framed in terms of protest or resistance, as they were concerns about the practicalities of implementation. When the talk of classroom applications arose, the teachers spoke openly about their individual experiences of teaching and the obstacles that they already perceived in educational practice – the vast majority of the which where rooted in resource limitations and time constraints.

In a conversation over coffee later that day, for instance, one of the teachers told me that the Ministry provided only 40 *colones* for lunch for each student – hardly enough to buy vegetables or a variety of foods – which is why the students were given rice, beans and soup every day.[5] *It's one thing to talk about the need to promote good health*, she said, *but without better support from the Ministry, how are we to achieve this?* Another teacher also confided privately that she did not believe that all teachers necessarily practiced the values that they were talking about promoting through the transversal themes: *How are the children going to learn if they do one thing in the classroom and another elsewhere? We talk about recycling all the time, but the school itself isn't recycling, so what sort of message does this send?*

The frustrations expressed by the teachers during this particular day in Santa Elena were strikingly similar to those that were shared with me by educators in other schools in Monteverde and elsewhere in Costa Rica over the course of the research. Many state educators I spoke to throughout the year expressed a genuine commitment to providing their students with both a strong values-oriented education as well as comprehensive knowledge of the basic subjects, but felt hampered in this effort by a lack of support from the state education bureaucracy. This was the case not only at the primary level in Monteverde, but also at the state secondary school. In addition, educators at the *Colegio* in Santa Elena also expressed concern about strong pressures from parents and local employers to provide older students with specific kinds of training or skill sets. In particular, the local tourism industry exerted a powerful, although often indirect, pressure on the school to give older students sufficient training in English and marketable knowledge of the local environment.

State Secondary Schooling

The state secondary school, *Colegio Técnico Profesional de Santa Elena*, is located along the main road, behind a tall wire mesh fence, on a hill above the commercial centre of Santa Elena. On this gently sloping property are a set of long, narrow, one-storey concrete breeze-block buildings painted white with blue tin roofs. Within the first of these buildings, the main office, teachers' lounge, computer lab and classrooms are arranged in a row, each with its door leading out onto a covered concrete walkway. A second long building runs parallel to this one, and holds more classrooms and the school library. Beyond these, and further from the main road, is a small concrete football pitch, a large area used for agricultural projects (including vegetables and coffee), a pasture with cattle, a large chicken house, and an artificial pond used for an experimental fish farming project.

---

[5] For purposes of comparison, at the time a litre of milk cost approximately 400 *colones* and a loaf of bread around 600 *colones*.

During school hours, the site was filled with the noisy presence of its several 100 students, ranging in age from 13 to 18. Many of these arrived early each morning on buses from smaller neighbouring communities, and many had travelled long distances to attend. These students may have boarded the bus in their home communities as early as 4.30 a.m. in order to arrive for the start of the school day at 7 a.m. Many students chose to make this long commute to Santa Elena because it was widely believed to offer better educational opportunities than secondary schools in other, smaller communities in the region. Students were required to wear the standard state school uniform, but the atmosphere of the school was relaxed, with teachers and students working together in classrooms and chatting in the corridors. Despite the generally positive atmosphere, however, the classrooms themselves were as uniformly empty as those in local state primary schools, and many contained only a chalkboard on one wall and a set of aged student desks. Also similarly to local primary schools, books, textbooks, and other kinds of learning materials were often in short supply.

The *Colegio* in Santa Elena was one of only four schools in Costa Rica in 2003 which offered a specialist programme in ecotourism (*turismo ecológico*).[6] The programme was first introduced on-site as a technical/vocational qualification in 2002. Prior to this, the school offered a specialisation in tourism (*turismo*) which focused on business management aspects of tourism such as accountancy and hotel management, and was intended to provide students with skills for future employment in local restaurants, hotels, and transport companies. The new ecotourism programme, in contrast, focused on the acquisition of knowledge about the local environment and tourism industry, national environmental history and legislation, and international environmental policy. Rather than seeking to train students to become future tourism business owners, the new programme was intended instead to train them to take up work for local conservation areas and other nature tourism destinations as nature guides, environmental educators or other protected area staff (e.g. guards, administrators, management).

As outlined in the previous chapter, for the first 3 years of secondary schooling (7th–9th grades), the national curriculum requires students to receive classes in the standard subjects. At the end of the ninth grade, students are allowed to choose a specialisation. For the last 2 years of secondary education, therefore, students carry a double course load composed of the basic subjects as well as courses in their specialisation. Significant class time is also spent during these later years in preparation for the final set of national exams in the basic subjects; these must be passed in order for the student to receive his/her *Bachillerato* (diploma). In addition, students must

---

[6]The school also offers a specialisation in textiles which, for reasons of space, I will not detail here. It is worth noting, however, that there is a noticeable gender divide evident in the choice of specialisation. The textiles specialisation is overwhelmingly taken up by female students, while both male and female students opt for ecotourism. Although I have not addressed gender issues here, they undoubtedly have impacts for individual students and teachers as well as the local tourism economy. For more on gender and schooling in Costa Rica, see the excellent work of anthropologists Karen Stocker (2005) and Ilse Abshagen Leitinger (1997).

pass a national exam for their specialisation in order to attain a *Titulo Técnico* (technical qualification). These multiple choice exams are almost entirely based on rote memorisation skills and mastery of a set of information specifically determined by the Ministry of Education. The national exam for the tourism specification in 2003, for example, was composed of 80 questions which required students to identify specific tourism destinations on a map, to match the names of mountain peaks to the protected areas in which they are located, and to match particular types of tourism (ecotourism, specialist tourism, etc.) with their defined characteristics, as well as to display knowledge of business management, national labour legislation, and computing skills.

Environmental Learning in the Ecotourism Programme

According to one teacher heavily involved in promotion of environmental learning at the *Colegio*, environmental learning was provided to students in a variety of ways. Firstly, it was supposed to be integrated into basic academic studies during all study years through promotion of the Ministry of Education's 'transversal themes' as well as individual teachers' creative approaches to teaching, especially in the natural sciences. Secondly, the school's teachers received support and extra programming from the environmental education co-ordinator at the Santa Elena Reserve (whose work is discussed in detail in Chap. 4). However, perhaps the heaviest concentration on these subjects was found within the ecotourism specialisation, for which the curriculum encompassed five broad subject areas: environmental education, environmental management, ecology, ecotourism, and English. Overall, the specialisation required intensive study of ecology (especially identification and memorisation of endemic species of flora and fauna), discussion of environmental concerns and management, and development of an understanding of national and international environmental legislation and the agencies involved in their promotion. Ministry of Education policy advised teachers to orient these discussions towards local ecology and local environmental concerns and management, with the understanding that the majority of students would seek work in their local tourism industry. Indeed, as the majority of Monteverde's population was either directly employed or indirectly involved in the tourism economy in some way, this seemed the most likely outcome.

In addition to the clear need for students to learn relevant content (not least as a result of the nature of national examinations), the curriculum documents for the specialisation were also strongly framed in terms of teaching concepts, values and ethics. This tension between attention to content and values mirrored that found in the wider national curriculum. In 2003, the 'environmental education' component of the tenth grade curriculum centred on:

> '… the development of students' interest in the management and control of situations that bring about environmental deterioration. Through an understanding of the causes and effects of specific concerns such as contamination, inadequate planning, exploitation of

resources, and the impacts of production activities, among others, it will stimulate the search for solutions which promote conservation of the environment and sustainable resource use... Students should be converted from simple observers to critical thinkers and protagonists of change.'

The environmental education curriculum outline that followed this statement was displayed in tables containing concepts, contents, suggested activities, and points for evaluation, that illustrate how teachers should manage the requirements. Concepts in the area included a broad definition of environmental education, as well as discussion about environmental ethics (*etíca ambiental*) and values (*valores ambientales para concientización ciuadana*). The curriculum document also suggested ways that these concepts could be put into practice, for example by requiring students to create environmental education projects for use in their own schools and communities (*proyectos para concientización del colegio y comunidad*).

Along with environmental education, in 2003 students in the tenth grade also studied three other components: 'environmental contamination', 'environmental legislation' and 'rural tourism'. The 'environmental contamination' section included detailed sections on scientific understandings of water, soil and air pollution, and was oriented around discussion of identified problems as well as their human causes and possible solutions. 'Environmental legislation', on the other hand, was an exceedingly heavy, overwhelmingly content-based section for which students were expected to become familiar with a long list of national laws, international treaties and conventions. The last section, 'rural tourism' was the most focused on social studies, and covered issues in rural economic development (*desarrollo economico del medio rural*) and 'rural culture' (*cultural rural y de los campesinos*), as well as environmental problems posed by agricultural production. As with the environmental education section, the curriculum document for all of these sections provided details on required concepts and contents, as well as suggested activities (e.g. group discussions, small group projects, posters and drawings, essays, presentations), and points for evaluation.

Teachers' Perspectives on the Curriculum

State schools teachers across Costa Rica are required to maintain – and be able to submit to the school director for inspection at any time – a notebook containing lesson plans that address each element of the required curriculum for each class they teach. Teachers often commented that such intensive paperwork demands made it difficult to cover all of the required curriculum material within the time limits either of individual school days or of the academic year as a whole. In 2003, the teachers' strike had made this even more difficult than usual. As Teresa Ramírez, a teacher in the ecotourism specialisation told me one day during a break between classes:

> *There is so much information to cover, it is really difficult. Time is always precious, but especially so this year because the teachers' strike means that we have about 6 weeks less of class time than usual, but we still have to cover all of the material so that the students can take their end-of-year exams.*

Teresa was born and raised in Monteverde, and was widely regarded within the school and the wider community as a highly dedicated educator. She told me that she saw her role as a teacher as fostering both the intellectual and the personal development of her students, and in order to do this she planned assignments that provided her students with time to think and work on their own. This was in line with Ministry of Education recommendations, she added, which encourage teachers to designate 40% of class time for communication between and among the students and the teacher:

> *The thing is, students can always sit down and read a book to learn the facts about something, but they can't learn values that way. Values can only be taught through conversation and interaction. If the students aren't being taught values at home, then it is even harder. I prioritise conversation and teaching values over the contents of the curriculum whenever I have the opportunity.*

Teresa also told me that she tried to organise extra projects and activities that would get her students involved in the wider community. While completing a unit of study on frogs, for example, she took them to visit the *Ranario* (known in English as The Frog Pond), a local educational ecotourism attraction. As part of previous studies, this group had also visited the local serpentarium and the Santa Elena Reserve. Finding the time and resources to do these kinds of trips can be problematic, however, she commented. Although all three of these businesses offered free entry for local students, each student still had to spend around 1,000 *colones* for transportation and food, and some simply could not afford to pay it.

Despite the financial limitations, Teresa and a number of other teachers at the *Colegio* devoted considerable time and energy to providing students with these kinds of learning opportunities outside the classroom. The school's director confirmed that for the six teachers working within the ecotourism specialisation that year, many of their most creative and inventive projects were organised because of their own individual interest in, and commitment to, environmental learning and community development. As such they were officially 'extracurricular' activities.

That teachers at the school devoted considerable time organising such additional learning opportunities for their students was all the more remarkable because the *Colegio*, like local state primary schools, faced on-going problems with a lack of teaching materials and high rates of teacher turnover. The *Colegio's* director told me that the school's most significant concern was that there were never enough resources to implement projects, despite the curriculum itself being well-formulated. Teresa also cited as one example of this the problems she had finding teaching materials for a unit on environmental law and organisations. The curriculum required her students to be able to understand, analyse, and critically assess a long list of domestic environmental laws and international treaties: *They have to at least have read them first. I've been able to find copies of all the laws except the Water Law, because it is under revision right now. I even looked at Universal and asked the AyA, but no one has it.*[7]

---

[7] *Universal* was one of the largest bookshops in the capital city; AyA (*Acueductos y Alcanterillados*) is the state water management agency.

She had also encountered problems in teaching about local ecology. The curriculum required students to become familiar with, and be able to identify, a number of locally, nationally and scientifically important birds, reptiles, amphibians, land mammals, insects and plants. However, the school did not have textbooks for either the students or the teachers to consult, and there was no public library in the region. Teresa instead relied on one field guide to teach about birds, another for reptiles, frogs, and insects, and yet another for plants. An additional frustration for many teachers in the local state system was that the great majority of written resources on local environmental topics were available only in English, if they were available at all. Such material limitations posed significant problems for teachers who had lived and taught in the region for a few years, and who already had some knowledge of the available resources, but could be even more daunting for those that were new to the region or to the profession.

Teachers at the *Colegio* in Santa Elena were in these ways consistently caught in a bind between meeting the requirements of the national assessment system, successfully addressing the explicitly values-based educational philosophy of national education policy, and also simultaneously providing the skills and knowledge required for students to gain access to employment in the local economy – all within the context of on-going financial and resource limitations within the state system.

The Cloudforest School

Despite the best efforts of local educators like Teresa to manage such concerns, the sometimes severe financial constraints experienced within state schools led many local parents to seek educational opportunities for their children within local private schools. These parents saw private education as providing distinct advantages over state school education, particularly in terms of schools that conducted classes in English. Private education undoubtedly did provide some practical advantages to both students and their teachers, but implementation of environmental education was just as problematic as it was in state schools. Like their state school counterparts, local private school teachers were heavily impacted by the curriculum content and national assessment requirements imposed by the state education bureaucracy. In addition, the social positioning of individual private schools within local and national contexts often served to further complicate the negotiation of definitions and practices of environmental education.

One particularly good example of the complicated social and economic relationships involved in environmental education in private schools is provided by the case of the Cloudforest School (also the *Centro de Educación Creativa*, or CEC), a local private school that describes itself as specialising in environmental education. According to the school's own historical account, it was established in 1991 by five local families who were concerned by overcrowding and what they perceived were the low educational standards of local state schools. The founding group included North American and Costa Rican settlers, members of the local Quaker community, and a locally-resident North American biologist (CEC 2002). It was at the suggestion of the biologist, who began doing research in the region in 1973 before permanently settling with his family in the village of Monteverde in 1980, that the school adopted its environmental education orientation. He told me during an interview in 2003:

'…. when it came time to establish the Creative Learning Center, for example… now called the Cloudforest School… it wasn't an effort by a bunch of biologists that did that. It was a reaction to educational opportunities in this area that led to that establishment. And I just happened to be sort of carried into it because my wife was at the original meeting, and as soon as I got my word in I said, "well, you know, let's make an environmental school out of this". Everybody thought… the particular group we had, a small group… thought that was a wonderful idea, you know, "Let's do it", even though most of them didn't have much of an idea what that might entail. But we, of course, eventually found out what that entailed and we're still working on it. We're still trying to make it happen the way we originally dreamed it.'

Classes were conducted in local homes for a small number of students until 1992, when a 42-hectare forest property was purchased with a loan from The Nature Conservancy (TNC). US biologist George Powell (who was also a central figure in the establishment of the Monteverde Reserve) was influential in convincing the TNC of the biological value of the virgin forest on the site, and helped to establish Costa Rica's first conservation easement on it. Since that time, the school has rapidly expanded with support from the non-profit Cloudforest School Foundation based in Tennessee (USA) as well as supplementary fundraising including individual programme grants (largely from US foundations), an annual sponsored walk-a-thon, proceeds from poster sales, and a 'plant-a-tree' donation programme. By 2003, the school's infrastructure had grown to include four sets of buildings for classroom and office space, hiking trails, ornamental gardens, organic vegetable and herb gardens, and a greenhouse, all situated on top of a hill outside of the centre of Santa Elena. The construction was completed in stages over a 10-year period, much of it with materials and labour donated by parents, students, school staff and other local supporters.

At the time of this research, the school's total enrolment had grown to include 173 students ranging in age from pre-school and primary school (1st–6th grade) through secondary school (7th–10th grade).[8] The majority of students – 60% according to the director – were receiving some kind of scholarship from the school, with the exact amount of support decided by a school committee on the basis of individual need. Approximately 90% of the student body during this time were Spanish-speaking Costa Ricans, while more than half of the teaching staff (16 out of a total of 25) were either experienced teachers or recently qualified teaching graduates recruited from the United States.[9] Basic subjects in the primary grades were

---

[8] The school planned to add an 11th grade when its oldest cohort had reached the appropriate age and enough students had been enrolled to make it reasonable to manage them as a separate group.

[9] National identity is often a problematic distinction locally because there are a number of local residents born into the community who are descended from the original North America Quaker settlers or from foreign researchers who have settled permanently in the community and who identify themselves primarily as Costa Ricans. In this case, I refer to students whose parents are Costa Rican nationals themselves and whose first language is Spanish.

conducted in English, with additional classes in Spanish (including grammar, vocabulary and Costa Rican history) each day. Secondary school students were also taught in a mixture of both Spanish and English, usually depending on the abilities of individual instructors. At the upper levels, however, teachers were under additional pressure to prepare students for the national state exams, which are administered solely in Spanish. The school's steady growth since its establishment was attributed both to the gradual addition of grades as the first classes of students matured, and also through the increased enrolment of new students after the school received accreditation from the Costa Rican government in 2002. According to the director, prior to the accreditation local Spanish-speaking parents tended to allow their children to attend for only a few years of primary school before moving them to one of the local state schools. This was because many parents wanted their children to learn English, but were also concerned that they received adequate preparation for national exams. Official recognition by the state, however, meant that more students were remaining in the school throughout primary and secondary levels.

The school's mission statement outlined its overall goals in the following way:

> '*Centro de Educación Creativa* is a bilingual, environmentally-oriented school in the multicultural cloud forest community of Monteverde/Santa Elena, Costa Rica. Students are mostly native Costa Ricans who will inherit responsibility for preserving the surrounding cloud forest and making sustainable development a reality. We intend to develop students' skills to do this effectively and, through a bilingual immersion approach, the voice to do so on a global scale. *Centro de Educación Creativa* promotes environmental awareness and responsibility by incorporating environmental education into every segment of its interdisciplinary curriculum. We will encourage the development of well-rounded persons by addressing the spiritual, mental and physical needs of each individual, while fostering an awareness and knowledge of community and world affairs and their affect on our environment.' (CEC 1995)

The mission statement provides an outline for an educational approach that is heavily child-centred and emphasises the importance of environmental learning. While many of the school's teachers, administrators, governing board and parents told me that they strongly supported this educational ethic, I also frequently witnessed intense discussions over curriculum content and the practicalities of implementing it. These disagreements can be linked to the complicated nature of relationships between people involved in the school and their sometimes conflicting definitions of environmental education, as well as to the school's relationships to the wider community and the state education bureaucracy.

Defining and Implementing Environmental Education

Perhaps one of the main reasons for these on-going negotiations about the content and orientation of environmental education at the Cloudforest School was that, as of 2003, the school still had no officially agreed definition for it. Although the founders provided a compelling justification for their choice to orient the school around environmental education (as evidenced in the mission statement above), the development of a curriculum which integrated environmental learning into every subject had proven more problematic. These difficulties can be attributed to a number of factors, including the school's rapid growth, its high rates of teacher turnover, and the diversity of perspectives on environmental education represented by individual teachers, staff, board members and parents.

When the school was first established in 1991, teaching was very informally organised and decisions about topics for study were largely left up to individual teachers. According to one teacher who had been at the school since its first year:

> 'When I first came here you came up with your own themes, often with what was interesting for the children, what their interests were, and that's what I would help guide them in… And then we found as the school grew that that wasn't enough for most teachers. They needed more structure.'

The school's first formal curriculum was created in 1994, under advice from a curriculum consultant from the US. It was based on the American Association for the Advancement of Science's *Benchmarks for Science Literacy* (1993), which contained guidelines for educators interested in improving scientific literacy in primary and secondary education.

The curriculum had subsequently been revised several times, and in November 2002 I was invited by the school's environmental education co-ordinator to participate in the most recent revision. This was carried out during a series of meetings held at the school over the course of the academic year. The committee was organised by the environmental education co-ordinator – a young woman from the US with experience of teaching both in classrooms and at an informal environmental education centre there – and also included the director and several classroom teachers (also all from the US). Although all students received some exposure to environmental education through weekly sessions offered by the environmental education co-ordinator and the school's land manager, at the time of these meetings the integrated environmental education curriculum which the school wanted to implement was only being used in a somewhat limited way and only in the primary grades. Teaching at the secondary level, on the other hand, operated on a more traditional course system with separate classes for each basic subject plus environmental education as an additional course. The purpose of the curriculum committee meetings, therefore, was to revise the existing curriculum in order to make environmental education a more integral part of classroom learning in all subjects and on all levels. As in the case of the national state curriculum, educators at the Cloudforest School hoped to be able to do this while also preparing students to succeed on the national exams.

For the primary grades, the curriculum being revised was a complicated system organised within three thematic strands: 6-week long topical themes, over-arching year-long ecological themes, and country themes. The 6-week themes included topics such as 'ecosystems', 'myself, my family, my community', 'early earth', 'democracy in Costa Rica', 'nutrition and reproduction' and 'simple machines'. After completing each 6-week long theme, students had 1 week of holidays before returning to start the next. Unlike the state system, in which the academic year runs from February to November, the Cloudforest School's year ran from mid-July until mid-June with 1 month-long break for the Christmas holiday and another between academic years. This timetable allowed for six themes per year, or fewer if individual teachers chose to extend one or combine others. Accompanying the 6-week themes were larger year-long themes that were intended to give teachers a guide for orienting classroom activities towards a specific ecological focus. These included 'awareness' (of people, animals and the environment), 'diversity and classification', 'systems and cycles' (solar, water, weather), 'land and people' and 'interrelationships'. Thirdly, country themes had previously been chosen for each grade – somewhat at random, members of the committee told me – as tools to raise awareness of human diversity. These included Costa Rica, Nigeria (as an example of social and cultural diversity), Australia (focusing on endemic species and adaptation), Japan (focusing on water issues), and Greece (the 'cradle of democracy').

As this rather complicated description suggests, the multiple overlapping themes and requirements had led to a good deal of confusion for classroom teachers, and particularly for those who were newly qualified or new to the school. Indeed, many teachers told me that they were uncertain of precisely what the curriculum requirements were for each year or how to meet them. Discussions during the curriculum committee's meetings revealed that group members also believed that the curriculum needed to be simplified and further formalised, both to assist current teaching staff and also because of the school's rapid rate of teacher turnover.

This high rate of turn-over – almost 50% of teaching staff in the previous year – was rooted in several factors. Firstly, many of the school's teachers were newly-qualified or had come to the school (most often from the US) because of a personal interest in environmental education. Few came with the intention of staying more than a few years in order to gain 'international' teaching experience or knowledge of environmental education. Additionally, the rate of pay at the school was relatively low in comparison to that of state school teachers, and the high cost of living in the community often made it difficult for teachers to stay for more than one or two academic years. This was particularly true of those with families, although there were many in this group who expressed a real interest in settling in the community long-term.

The majority of teachers and administrative staff I spoke with believed that this high rate of turnover had significant implications for the school's ability to achieve the sought-after integration of environmental education into all areas of the curriculum. While some teachers had been able to effectively integrate environmental issues into classroom teaching because they had significant experience working with these topics either at the Cloudforest School or elsewhere, others commented that they had little or no previous training in environmental education topics or methods. As a result, they tended to rely on the school's environmental education co-ordinator and land manager to teach environmental education topics during their designated weekly sessions. Still other teachers spoke of the pressures that they faced in preparing students for national exams and their frustration at therefore not having sufficient time to 'integrate' environmental education, or about the inherent difficulties of integrating environmental topics into subject areas such as literature or mathematics.

One of the other central areas of discussion at the first curriculum committee meeting in November 2002 was the school's continuing lack of an official definition of environmental education. This had led to a variety of understandings and approaches to environmental learning by those involved in implementing it, including classroom teachers, the environmental education co-ordinator, and the land manager. The lack of an agreed definition, along with confusion over curriculum content, had led many teachers to instead organise their classrooms according to their own individual interest in or commitment to certain topics, ways of teaching or goals for environmental learning. The range of instructional activities used at the school thus also ranged widely, from teaching taxonomy in the classroom to taking walks in the forest, working in gardens (described to me as a 'hands in the dirt' approach), or completing community service projects.

One teacher, for example, who had taught at the school for 10 years since immigrating from the US, defined environmental education as an opportunity for students to have a 'magical' experience in which they learn to feel connected to the natural world. As we sat at a picnic table watching her students play outside one morning, she told me:

> 'What we feel is important is the *concepts* of science, of social studies, those kinds of things, rather than just information. Not "how many miles long is the river?" but "where does it come from?", "where does it go?", "what is the cycle of water?".... those kinds of things... it's about understanding the cycles [of life] but making it magical.... having imagination play a part in it.
>
> So, we have, for example, in first grade, a semester-long activity where we are visited by an imaginary dragon. We never see him. But the dragon has a journal. He's looking for children who love the earth, and he sees that things aren't being real well taken care of. And he wants to find children who love the earth and love the stories of the earth. So he leaves us a map. It's done in a sense of excitement. We see this letter from a dragon... it's kind of burnt around the edges like maybe a dragon would have...[laughter] And we go out and we find his box. What are his treasures? Things like a feather, a smooth stone, things that he feels are special. So then that starts us off on the stories about the earth, maybe creation myths, native stories here of Costa Rica, different tribes.
>
> We also do things that are sensory-awareness for development. Like with the dragon, one of the journal entries is a walk that he takes. The dragon – he or she describes this walk and we take the journal and try to trace that path. So he says: "I saw this tree, it was huge, it was enormous" and he was describing a strangler fig tree. So we find that tree. And he says, "I found this smooth red rock and I left it under the tree." And sure enough there's the rock.
>
> So it's fun in that sense that it's magical, but at the same time... as they do this, they earn different beads when they can show an understanding of different concepts. So for the 'inter-relatedness of life', we have activities and things, where they can show me whether they can speak English or not, and whether they can write or not. Maybe it's through dramatisation, maybe it's through pictures or posters or whatever that they made, maybe it's through a story that they can tell – if it's in English or in Spanish – that shows that they understand the concepts.'

This approach contrasted strikingly, however, with the definitions and goals for environmental education offered by other educators at the school. The school's land manager, José, a Spanish-speaking Costa Rican from the neighbouring community of San Luis, for instance, provided weekly lessons for classes in each grade level, but his interactions with students centred on teaching them practical skills such as how to germinate plant and tree seedlings, to manage a garden, or to make compost. Rather than taking place in a classroom, José's lessons were located in the school greenhouse, gardens, or forested areas, with the learning activities designed to be appropriate to the age and knowledge-level of each student group. Younger students were given responsibility for practical small-scale projects such as collecting decaying plant material from the forest floor to serve as an organic base for seed germination. Older students, on the other hand, were responsible for overseeing reforestation efforts on degraded areas of the school property. Interestingly, José argued that what he did was not 'environmental education', or at least not in the same sense that others at the school defined it:

> *The environmental education co-ordinator deals with all the theoretical ideas and the curriculum requirements. What I do is the **practice**. I believe that this is the most important part for children to learn – conservation, reforestation, recycling, rescuing plants and, very importantly, soil conservation.*

This diversity of ideas and practices – sometimes resulting in conflict and sometimes in collaboration between educators – is an illustration of the multiple tensions that can surround environmental education and learning in practice. In the case of the Cloudforest School, these tensions played out not only within the school itself, but also in its relationships to the community and the state. This is because, like their state school counterparts, educators at the Cloudforest School faced limitations on what they believed was ideal educational practice because of the influence of the state education bureaucracy as well as the impacts of local economic and social circumstances.

Relationships to the State and the Community

In concrete terms, the most serious restriction which educators at the Cloudforest School faced in the implementation of the desired integrated environmental education curriculum was due to the strict curriculum content and assessment requirements of the Ministry of Education. Interestingly, while educators at the school described their curriculum as significantly different from the one used in state schools, my own review suggested that its content was in reality quite similar. This is likely because in order to progress through the relevant years of schooling and to advance into higher education, the school's students had to pass the same content-based national examinations as their state school counterparts. The school was forced to accept these curriculum and examination requirements in exchange for its accreditation from the state. Although accreditation had meant that more students stayed on long-term at the school, teachers (and especially those working with older age groups) complained that the demands of preparing students for national exams often left little or no time for any additional or creative teaching or activities. This was even true of environmental education, many teachers told me, despite the school's stated mission to create environmentally aware students. As a result, discussions amongst staff and board members at the school frequently centred on the need to balance the teaching of environmental education concepts with sufficiently preparing students for exams. According to the director:

> 'The tests really constrain us quite a bit. There's testing in sixth, ninth and eleventh grade, so luckily the elementary is pretty free until sixth grade. There are some topics we touch on in other grades that will be on the exam. But then in sixth grade it's like, "OK, we need to teach some of this curriculum" because the test is… it's not like an ITBS test that tests whether you can read and write and do basic math. It actually tests *content*. Science is taught that way too. It's a specific set of facts they need to know. So it's not conceptual, which is hard because that's really contrary to our educational beliefs. We really try to teach the "deep concepts" and not just memorising facts.'[10]

The state system's focus on teaching content was also strongly criticised by other educators at the Cloudforest School, and especially by teachers and staff who saw their school's approach to education as essentially transformative rather than content-based. As one teacher noted:

> 'The educational programmes here [in Costa Rica] are very much… you copy off the board, and that's how you teach something. In fact, I was at one of these workshops with other educators from the region, and the speaker was telling them about how important the students' notebooks are. Those are the official record of what was taught, so every lesson should have the date on the top and a lesson plan of what is being taught that day, before their notes are in there. And that was the *official record* of what happened that day. And we were just like, "Well, we don't really have those… Not everything we do is in the notebook". You know, [whispered] "We do stuff that's not in the notebook." It's a very different way of thinking about education.'

Such perceived differences in perspective on the role of education and its best practice had significant impacts on relationships between local schools, community groups, individuals, and state agencies, especially as they worked to implement environmental education either within one organisation or in collaboration with others. In some cases, however, perception of these different perspectives seemed to be as much the result of a lack of communication between educators as it was a statement about actual differences in educational practice. Many private school teachers in the community, for example, were largely unaware of the strong transformative element of the state's educational philosophy and environmental education curriculum, either because of linguistic and social divisions within the community or their short-term residence in the region. Equally, educators in state schools – as well as other community members – tended to categorise the Cloudforest School as an elite institution for 'gringo' children, despite the fact that the majority of its students were actually from local, Spanish-speaking, Costa Rican families.

This perception of the school had particularly noticeable impacts on its relationships with other types of community organisations and individuals. Whereas local state schools had strong relationships with the environmental educators at the Monteverde and Santa Elena Reserves (the subject of the next chapter), the common perception that

---

[10] The Iowa Test of Basic Skills (ITBS) is a standardised testing system used throughout the United States.

the Cloudforest School was well-funded and supplied with highly trained teachers had resulted in relatively little collaboration with other local organisations.

The director and many of the school's teachers also often expressed frustration that the parents of many Costa Rican students did not participate in school events and activities such as informational meetings or parent workdays (when parents were called upon to volunteer their labour to clean up the school grounds, paint, repair roofs, garden, or do other chores). When I attended one of these parent workdays in January 2003, I noticed that the vast majority of parents present were indeed foreign nationals (mainly from the US), many of whom were already heavily involved in the school through membership on the school board or involvement in fundraising activities. A few weeks later, one of the board members – a native Spanish and fluent English speaker – told me that part of the problem with getting parents to participate was that so many of the school's events were conducted only in English: 'Even the board meetings are held only in English. I can understand why it's that way, but it's so exhausting when it's your second language. It's really no wonder none of the parents want to get involved.'

Furthermore, despite the school's active promotion of its environmental education programme, many local parents sent their children to the school for other reasons. An informal survey of parents conducted in 1995, for example, found that an overwhelming majority sent their children to the school for the primary purpose of learning English, and considered environmental education programming a secondary or even lower priority (Dickey 1995). Fluency in English is of considerable importance to local employment, particularly in the tourism industry, so it is perhaps understandable that some parents prioritised language learning for their children. However, this revelation came as a huge disappointment to teachers, staff and school board members who saw the main goal of the school as promoting environmental learning in the community.

This emphasis on environmental learning was partly rooted in very strong relationships with local scientists and conservationists, many of whom were heavily involved at the Cloudforest School, usually either as the parents of enrolled students and/or as members of the school board. While environmental protection was considered important in the community generally, local scientists and conservationists have historically been the strongest voices in debates about local conservation efforts, and have argued in the main for strictly protectionist modes of conservation. Their perspectives on conservation have also been influential in the negotiation of the content and aims of environmental education at the school.

Teaching Science or Cultivating Values?

In particular, in 2003 there was considerable pressure to orient environmental education at the Cloudforest School towards teaching in the natural sciences. This type of approach to environmental learning privileges scientific understandings of environmental management arising out of research in biology, botany, and ecology.

Proponents argue that once children are taught to understand the natural world (through learning in the natural sciences), they will come to love it and work to protect it. Underlying this approach is an assumption – frequently critiqued by both practitioners and theorists of environmental education – that exposure to certain kinds of (scientific) knowledge will have a directly transformative effect on students.

This understanding of the nature of environmental learning stands in stark contrast to those – endorsed by other individuals and organisations in the community – which focused more heavily on social learning related to the environment. These approaches give topics such as equality, justice and responsibility a more central role in learning about environmental topics and issues, and often use a very different range of teaching and learning strategies. As the next chapter will discuss, the contrasting understandings of environmental learning in Monteverde are linked to very different agendas for environmental management and community development in the community.

The case of these diverse schools also highlights that it is not only at the international level that issues around education and environmental knowledge are debated, practiced, defined and re-defined. It is commonly acknowledged that schools are important sites for knowledge transmission, the negotiation of meanings and priorities, and processes of economic and social development. Certainly, all of Monteverde's schools were embedded in social and economic relationships which had a significant impact upon decisions about curriculum content and teaching practices. Interestingly, despite their differences, the schools shared a number of common challenges to their environmental education efforts, including limitations imposed by the state education bureaucracy (and especially the demands of the national assessment system), on-going problems with implementation due to a lack of teaching and financial resources, and the demands of parents and local employers.

It was partly in response to these limitations that both state and private schools in Monteverde have worked to develop supportive relationships with local organisations. Educators in state schools in 2003, for example, relied heavily on local conservation groups to provide additional programming and resources which were not available from the Ministry of Education. Private schools' relationships with these groups tended to be more limited for a variety of social and economic reasons, but local private school teachers and students also took advantage of learning opportunities within local protected areas, tourism destinations, or projects funded by local organisations. Local schools were thus part of extensive social and economic networks which connected them to other kinds of local organisations. It is these relationships, and their impacts on environmental education in schools and in the wider community, that are the subject of the next chapter.

References

American Association for the Advancement of Science. (1993). *Benchmarks for science literacy*. New York: Oxford University Press.
Aylward, B., Allen, K., Echeverría, J., & Tosi, J. (1996). Sustainable ecotourism in Costa Rica: The Monteverde Cloud Forest Reserve. *Biodiversity and Conservation, 5*, 315–343.

References

Bloom, B. S. (1984). *Taxonomy of educational objectives*. Boston: Allyn and Bacon.

Blum, N. (2008). Ethnography and environmental education: Understanding the relationships between schools and communities in Costa Rica. *Ethnography and Education, 3*(1), 33–48.

Burlingame, L. (2000). Conservation in the Monteverde Zone: Contributions of conservation organisations. In N. Nadkarni & N. Wheelwright (Eds.), *Monteverde: Ecology and conservation of a tropical cloud forest* (pp. 351–388). Oxford: Oxford University Press.

CEC [Centro de Educación Creativa/Cloudforest School]. (2002). *History of the Cloud Forest School*. Unpublished internal document. Monteverde, Costa Rica: The Cloudforest School.

CEC. (1995). *Eight Year Plan 1995–2003*. Unpublished internal document. Monteverde, Costa Rica: The Cloudforest School.

Dickey, K. (1995). *Parents' assessment of the Creative Learning Center, a growing private school in Santa Elena, Costa Rica*. Unpublished paper for the Tropical Field Research Program, Associated Colleges of the Midwest.

Dutschke, M., & Michaelowa, A. (2000). Climate cooperation as development policy: The case of Costa Rica. *International Journal of Sustainable Development, 3*(1), 63–94.

González Gaudiano, E. (2007). Schooling and environment in Latin America in the third millennium. *Environmental Education Research, 13*(2), 155–169.

Leitinger, I. A. (Ed.). (1997). *The Costa Rican women's movement: A reader*. Pittsburgh: University of Pittsburgh Press.

Little, A. W., Hoppers, W., & Gardner, R. (Eds.). (1994). *Beyond Jomtien: Implementing primary education for all*. London: Macmillan Press.

MEP [Ministerio de Educación Publica]. (2002). *Los temas transversales en el trabajo de aula*. Curriculum document. San José, Costa Rica: Ministerio de Educación Publica.

Monteverde Friends Meeting. (2001). *Monteverde Jubilee Family Album*. Monteverde, Costa Rica: Monteverde Friends Meeting.

MVI [Monteverde Institute]. (2002). *Encuesta de desarrollo: Encuesta para residencias*. Unpublished survey data from the Monteverde Institute and the USF Globalization Research Center. Monteverde, Costa Rica: Monteverde Institute.

Nussbaum, M. (2000). *Women and human development: The capabilities approach*. Cambridge: Cambridge University Press.

Sen, A. (1999). *Development as freedom*. Oxford: Oxford University Press.

Schultz, T. W. (1961). Investment in human capital. *The American Economic Review, 51*(1), 1–17.

Stocker, K. (2005). *'I won't stay Indian, I'll keep studying': Race, place, and discrimination in a Costa Rican high school*. Boulder: University Press of Colorado.

UNESCO. (2000). *The Dakar framework for action*. Paris: UNESCO.

Chapter 4
Environmental Education and Conservation Organisations*

Abstract One of the key on-going debates in environmental education research and practice relates to the content and goals of programmes. Specifically, there is a long history of debate between advocates of educational perspectives that emphasise the teaching of science concepts and those that seek to more actively link environmental and social issues. In practice, educators and organisations respond to these tensions in a variety of ways, often due to the particular social and economic contexts in which they are located. This chapter explores these debates about the 'appropriate' content and aims of programmes by looking at the case of environmental educators working within two conservation organisations in Monteverde, Costa Rica. It reveals that environmental education (i) is an important local site of debate about understandings of the natural world and humans' relationships to it, and (ii) is part of much wider struggles over the control of processes of local development and environmental management.

Keywords Conservation organisations • Environmental education • Science education • Transformative learning

Alongside local schools, NGOs of various types form another important part of the network of sources for environmental teaching and learning in Monteverde. The vast majority of local NGOs are oriented around environmental issues, either directly (through land management and conservation activities) or indirectly (through involvement in the local tourism and service sectors). The local women's craft co-operative, CASEM, for example, made a strategic decision a number of years ago to use environmental designs and imagery in its products in order to attract tourists (Leitinger 1997a). Conservation NGOs have been the most highly visible

---
*A version of this chapter was previously published as Blum (2009).

and influential organisations in local decision-making since the community was founded, and were also the very first local organisations to begin promoting environmental learning in the region.[1] Indeed, several local groups had already begun working in environmental education prior to its inclusion in the national curriculum, and in 2003 programmes organised by local conservation groups continued to support and enhance the work of educators in the state school system.

Educators working within local conservation organisations were situated somewhat differently in relation to issues of educational content and pedagogy than their colleagues in the formal sector, however. In contrast to state school educators, those working for conservation organisations operated in relative isolation from state education bureaucracy and as a result they often had a wider scope for implementing programmes in accordance with their individual perspectives on environmental education and beliefs about the ways in which it should be used to promote change in the community. At the same time, these educators were also under pressure from their employers to implement programmes that would enhance broader organisational goals and interests. These goals varied widely depending on the particular financial, institutional and ideological interests of each organisation. Groups variously saw their primary role as promoting the strict protection of local forests, encouraging scientific research, or acting as a leader in community development. Each group's financial circumstances were also varied, ranging from having easy access to international funding to relying heavily on support from the Costa Rican state. Educators thus negotiated definitions of environmental education and decisions about programme content in the context of both opportunities and limitations that resulted from their association with particular local organisations.

The diversity of organisational goals expressed by local conservation groups in Monteverde also highlights the ways in which environmental education was tied to broader community relationships and debates about the best way in which to achieve sustainable local development. While some organisations argued for strict forest protection and the promotion of scientific research, others argued for the promotion of the local tourism industry and other livelihoods activities that could provide some level of environmental protection as well as improving local standards of living.

To explore these issues 'on the ground', this chapter provides an overview of conservation NGOs and their activities in the community in 2003, as well as an account of the work of two influential local conservation NGOs – the Monteverde and Santa Elena Reserves – and the differing perspectives of their respective environmental education co-ordinators – Maria Rodriguez and Luis Delgado. The choice to focus on these two organisations is based on a number of factors. Firstly, despite often working collaboratively, the two organisations and their respective educators supported almost diametrically opposed perspectives on environmental education. This difference was rooted in the educators' respective personal experiences, training and commitments, as well as in the goals of the organisations for which they worked and the relationships of each organisation to the community as a whole.

---

[1] For a history of conservation NGOs in the community, see the edited volumes by Nadkarni and Wheelwright (2000) and Monteverde Friends Meeting (2001).

Maria was employed by the Monteverde Reserve, an organisation whose main goal is to provide strict protection of a vast forested area and to promote scientific research, and her programmes focused largely on the promotion of scientific understandings of local conservation strategies and environmental problems. Luis, on the other hand, was employed by the Santa Elena Reserve, a project that self-consciously labels itself as a community development project, and his programmes provided a style of environmental education that is firmly rooted in a commitment to social transformation and consciousness-raising. Their divergent perspectives and practices of environmental education serve as a useful frame for understanding local-level negotiations of programme content and of the concept of environmental education itself.

I also chose to centre the discussion around these two educators because of the extensive amount of time I spent with each one, their willingness to assist in the research, and their candid responses in our discussions. In both cases, the formal interviews I initially conducted with each of them gradually developed into collaborations that lasted throughout the year I spent in the community. These collaborations were based on informal arrangements in which I offered to assist each co-ordinator with their programmes in exchange for opportunities to observe sessions and communicate with participants. In some cases, this took the form of lesson planning or preparation of materials, and in others it involved overseeing student group exercises or contributing to group discussions during sessions. As the year progressed, Maria and Luis also invited me to accompany them to a variety of local gatherings – such as teacher training sessions or meetings with fellow educators – in which I could both observe and take part in discussions. The following chapter, therefore, is based on material I gathered both as a participant in their environmental education programmes and other activities, and from the many discussions I had with both co-ordinators over the course of the year.

Early Settlement and the Growth of Local Conservation

To begin, it is useful to provide an overview of the development of conservation efforts in Monteverde as well as some of the key issues in local debate and discussion.

Accounts of the early settlement of the Monteverde region state that the first settlers were a few Costa Rican families who made a living through subsistence farming of corn, beans, vegetables, fruits and livestock, and established two small schools and a few small stores. Little written history exists for this early era, although descendents of the first families recount that the first settlers arrived in the 1920s and 1930s (Griffith et al. 2000: 391). Limited road infrastructure made commercial agriculture relatively difficult, although a few upland farms did produce garlic, flax, beef and homestead cheese. Interest in dairy and beef production increased in the 1950s and 1960s as the main road to the capital was gradually improved. In 1951 a group of Quakers from the US state of Alabama arrived to set up their own settlement which soon included homes, farms, and a school. The nine families (25 individuals) had left the United States after several of their members were jailed for refusing to register for military service during the Korean War. In a published memoir of the first 50 years of the settlement, the 'Monteverde Jubilee Family Album' (cited hereafter as Monteverde Friends Meeting 2001), members of the original group of settlers recall that it was Costa Rica's reputation for peace and democracy that attracted them, as did the opportunity to purchase land for farming. After searching for suitable land in several other locations, the families settled on the region which they subsequently named Monteverde (literally translated from the Spanish as 'Green Mountain'). Members of the group also recount stories of the early 'pioneer' days when they lived in make-shift housing and began to set up new homestead farms. In all, the settlers purchased 3,000 acres of land, setting aside 1,000 acres for watershed and dividing the rest among the families (Monteverde Friends Meeting 2001: 16).

By the early 1960s, tropical biologists sponsored by the Organisation for Tropical Studies (OTS) began arriving in the region to study local flora and fauna, and soon after identified the Golden Toad (*Sapo Dorado*), an extremely rare endemic species. The discovery made the region famous within scientific circles and heightened research interest in the area's rich biodiversity. According to US biologist George Powell, when he arrived with three others in 1971 to do a study of army ants, he quickly became alarmed at the rate with which local forests were being logged and took the first steps to purchase the most biologically-sensitive areas for protection (Monteverde Friends Meeting 2001: 172). Powell worked with residents to motivate local support for the project, and found further support from the Tropical Science Center (TSC; *Centro Científico Tropical*), a consortium of Costa Rican and international universities and researchers based in San José, which gave the new reserve a formal legal status. The Quaker community was supportive of the reserve's creation and agreed to lease its 1,000-acre watershed to the TSC in perpetuity. A large land grant was also agreed with the Quacimal Land Company, and subsequent land purchase proceeded rapidly with the help of donations from international sources including the World Wildlife Fund, the Explorer's Club (New York), the Philadelphia Conservation Society, the International Council for Bird Preservation, the New York Zoological Society, RARE and a number of individual philanthropists (Tosi no date: 2). The Monteverde Cloud Forest Reserve (*Reserva Biologica Bosque Nuboso Monteverde*) officially opened in 1972.

In the early years, the reserve's staff – many of them volunteers – remained small, so attempts to communicate with local residents were largely informal, and often stimulated by concerns over land invasion by squatters or hunters. Visitation at the reserve was also initially limited to small numbers of scientific researchers – the majority of them coming from the US – so Monteverde Reserve supporters actively courted the interest of 'birders and TV producers, the only active "ecotourists" at the time' (Monteverde Friends Meeting 2001: 171), in order to raise the international profile of the newly created reserve and bring more scientists and students to the region to conduct research. Visitors were hospitably received by members of the nearby Quaker community, many whom lodged with families and stayed in the region for extended periods of study. A few settled permanently, while others became part-year residents or frequent visitors. When the Golden Toad disappeared after a still-unexplained population crash in 1987, scientific and conservationist interest in the region intensified even further. According to one of the early Quaker settlers, 'We had lived in Monteverde about 10 years when the first biologists arrived, intent on doing an in-depth study of the army ants. The scientists have been coming in greater and greater numbers ever since, until it would seem there is no bird, bat, butterfly or bug that has anything left to hide' (Monteverde Friends Meeting 2001: 168).

Throughout the 1970s and 1980s, the local population continued to grow as both Costa Rican and foreign settlers were drawn to the area's beauty. The community's infrastructure also grew significantly during this period with the establishment of a women's craft co-operative, a community arts centre, and two private schools. In addition, as the local tourism industry grew so did the variety of local businesses and available services, including hotels, art galleries, transportation services, supermarkets, restaurants, hardware stores, souvenir shops, laundries, tourist information centres and a regional post office. Local residents also began to open forest canopy tours, hiking trails, riding stables, and other kinds of ecotourism venues to cater to visitors' interests.

In 1985, a second local conservation organisation was formed by a group of largely US residents including biologists and members of the Quaker community. The Monteverde Conservation League's (known locally as '*La Liga*' or 'the League') original goals were to preserve forested areas on the Pacific slope of the Continental Divide (which the area straddles), to collaborate with the Monteverde Reserve's work on the Caribbean slope, and to promote research and education. In 1986, the group received its first small grant from the World Wildlife Fund for land purchase and it began an intensive land purchase campaign that continued until 1992.

A non-profit educational institution, the Monteverde Institute (known locally as 'the Institute'), was founded in 1986 as the result of 'a growing "love of nature" among Quaker families who welcomed the presence of biological researchers, and the visits of graduate tropical biology courses run by the OTS' (Trostle 1990). Two residents with experience in study abroad programmes, Quaker resident John Campbell and University of California at Santa Barbara biology professor Nalini Nadkarni, worked together with other local residents and contacts in the United States to establish an exchange programme with UC-Santa Barbara. By 1987, the first group of undergraduate tropical biology students had arrived, and in the first 5 years of its existence the Institute hosted more than 34 groups with approximately 528 participants (Monteverde Friends Meeting 2001: 206). This promotion of 'student' or 'scientific' tourism was considered by some community members as the best way to creatively manage a style of tourism in the region that would give quality educational experiences to visitors, while also creating job opportunities for local residents.

Around this same time, a third protected area was established by the local state secondary school (*Colegio Técnico Profesional de Santa Elena*; see also Chap. 3) and a group of Costa Rican residents. The small (310 ha) Santa Elena Reserve (*Reserva Bosque Nuboso Santa Elena*) is situated approximately 5 km outside of the village of the same name, and about 12 km from the Monteverde Reserve. The property is a mixture of approximately 80% primary growth and 20% secondary growth forest, and it is the highest elevation reserve (1,700 m) within the local reserve complex. The land is owned by the state, but was leased to the *Colegio* by the national government in 1983 when a project was established to provide the school's alumni with land for agriculture. The soils proved too infertile and the forest too difficult to clear, however, and the project was eventually abandoned. The government lease was maintained, however, and in 1992 the school along with a group of local residents decided to develop the property as a community-run ecotourism project, the profits from which could be used to support the financially-strapped school. The project's founders also envisioned using the property as a resource for student learning and training. Since that time, with assistance from various members of the community and Youth Challenge International, the property has been developed with 12 km of trails and a visitor's centre (see Wearing 1993; Wearing and Larsen 1996).

Prior to the establishment of the Santa Elena Reserve, some residents of the village of Santa Elena were reportedly resentful of the popularity of the Monteverde Reserve, because they believed that its profits were being sent to the Tropical Science Center in San José rather than benefiting neighbouring communities. The project's promoters and residents of the village of Santa Elena hoped that – in addition to benefiting the *Colegio* – the new reserve would also redress this perceived inequality and provide more jobs for community members. For its part, the administration of the Monteverde Reserve supported the establishment of the Santa Elena Reserve because of its potential to receive its over-flow visitors; for reasons of preservation visitation at the Monteverde Reserve is limited to 150 individuals on the trails at any one time (Aylward et al. 1996). Local business owners, concerned about overcrowding at the Monteverde Reserve and the possible future impacts on the local tourism industry, were also supportive of the establishment of this additional local tourism attraction.

In the years since their establishment, the Monteverde and Santa Elena Reserves, the Monteverde Conservation League and the Monteverde Institute have grown significantly in terms of membership, land ownership, and international recognition. By 2003, the Monteverde Reserve extended over 17,000 ha of mostly primary growth cloud forest, with the original Quaker watershed still at its centre. Following a massive international land purchase campaign, the League owned more than 18,000 ha by 1998, most of it within the internationally-recognised Children's Eternal Rainforest (*El Bosque Eterno de los Niños*).[2] In 1997, the Institute constructed a new 'ecologically-sensitive' administrative building in the village of Monteverde which includes office space, meeting space, and a library. In 2002, the organisation also purchased a 27 ha property overlooking the Gulf of Nicoya where it established a biological station for use in education and research, and was actively

---

[2] Since the time of this research, the League has continued to expand its holdings (see http://mclus.org/land-purchase-and-protection/).

engaged in a campaign to encourage landowners with properties located between the Monteverde Reserve and the Children's Eternal Rainforest to place forest fragments under conservation easements.

The establishment of new organisations, committees, commissions and other kinds of interests groups has historically been, and in 2003 continued to be, a characteristic part of community life. Groups that were formally constituted, such as the organisations mentioned above, co-ordinated a wide variety and large number of projects and programmes in conservation, research and education. Less formal groups, locally labelled as 'committees' or 'commissions', tended to have shorter life spans, and were commonly established to deal with a particular problem or to co-ordinate individual projects and events. Local tourism business owners also organised themselves both in formal terms (through business associations and the local Chamber of Tourism) and informal terms (often through influential family links) in order to promote the region and sustain the industry. Local residents frequently moved between groups, or maintained active membership in several simultaneously, depending on their interests or training.

This complicated organisational network provides opportunities for both collaboration and conflict in the community. In 2003, for example, the Monteverde Reserve provided guards and maintenance staff for a portion of the League's properties, while the League's Children's Eternal Rainforest served as a buffer zone for its neighbouring Monteverde Reserve. At the same time, a long-running legal battle between the League and the Monteverde Reserve (more specifically, the Tropical Science Center) over the right to protect between 4,000 to 5,000 ha in the Peñas Blancas valley continued to strain the relationship between the two groups during the time of this research (Griffith et al. 2000: 359). The League has also come under some criticism from local residents for its treatment of local landowners during its first land purchase campaign in the 1980s, as well as for particular problems with its San Gerardo project (see Vivanco 2006).

Science, Conservation and Local Environmental Management

One of the most significant points for local debate related to the number of diverse perspectives on conservation. While some individuals and groups supported largely science/research oriented approaches which involved setting strict limits on human activity in biologically valuable areas (sometimes known as 'fortress conservation'), others argued for styles of conservation which allowed for multiple-use strategies in order to more directly support local livelihoods. This is a common dispute in many areas of the world, and due to the economic importance of research and ecotourism to Monteverde's economy, these debates have taken a particularly central position in local politics and organisation.

As in the nation as a whole, scientific research and conservation organisations have played an important role in public debate and decision-making in Monteverde. As detailed above, when individual researchers began arriving from the US in the

1970s to study the area's rare and rich flora and fauna, many ended up settling permanently in the region.[3] Some of these early arrivals were influential in the founding of local conservation groups, including the Monteverde Conservation League, the Monteverde Institute, and the Monteverde Reserve. These local groups have succeeded in gaining financial support from international conservation groups, but the strong international reputations of many of Monteverde's researchers – several of whom are equally well-known and influential at the national level – and the community's own renown as a site of successful forest conservation has allowed local groups to retain a high degree of autonomy in managing local protected areas and conservation projects. As one long-time resident of the village of Monteverde, a biologist from the US, told me during an interview in 2003:

> 'None of those organisations have projects in Monteverde.... as you may know, the Nature Conservancy bought the land that the CEC [*Centro de Educación Creativa*] is located on. They bought that for them. And the World Wildlife Fund was very helpful in the campaign to establish the Children's Rainforest. It wasn't their project though. It was the League's project... I think they [international conservation NGOs] are probably are on top of the situation enough to realise that their efforts are needed more elsewhere in Costa Rica than here. Because the Conservation League may not be perfect, but hey look what we've already done and what we're doing, you know? Certainly a lot more than happens in most places. So why invest effort here, when you can invest it more profitably somewhere else?'

Despite the well-publicised success of many conservation efforts in the region, however, local researchers and conservationists continued to campaign for the extension of protection to further tracts of forested land. Existing groups have worked to expand their land holdings, and the newest local conservation group to be established during the time of this research – the Costa Rican Conservation Foundation (*Fundación Conservaciónista Costarricense*) – was founded in 2000 for the explicit purpose of buying land for protection on the Pacific Slope. This group, spearheaded by two local Costa Rican tourism business owners, a long-term resident biologist from the US, and the owner of a popular hotel in Santa Elena, argued that the majority of the land under protection in the region was on the Atlantic Slope and that this could have significant impacts on endemic species preservation. Indeed, research at the time suggested that more than 50% of the tree species found in forest fragments on the Pacific Slope did not occur within existing protected areas on the Atlantic Slope (Wheelwright 2000: 426). Pacific Slope habitats are also important to high-altitude migrants such as the Three-Wattled Bellbird (locally, *Pájaro Campana*) and the Resplendent Quetzal (*Quetzal*) – species that are locally significant because of their rarity, and the accompanying interest of researchers, bird watchers and other tourists. Pacific Slope lands are largely privately owned, however, and many areas have been in use for agricultural production for the last

---

[3] Costa Rican researchers, on the other hand, were less commonly present in the region during the time of this research, largely because the state had relatively few resources to support scientific research, even within the national system of protected areas.

several decades. Rapidly rising land prices, commonly attributed to the development of the local tourism industry, have also made land purchase and protection in these areas difficult.

During the many discussions that I had with educators and community members throughout the year, tensions were clearly highlighted between this protectionist agenda – which has overwhelmingly characterised conservation efforts in the region since the 1970s – and alternative perspectives on community development. These debates mirror the conflicts between the 'traditional conservation narrative' (strict protection) and 'conservation counter-narrative' (multiple-use; sustainability) which have also been identified at the national level in Costa Rica (see Campbell 2002). More specifically, local debates centred around understandings of the relationships between environmental protection and 'development', especially in terms of the need for improvements to local infrastructure and communications, as well as the more recent emergence of discussion about 'sustainable' or 'integrated' community development.

In 2003, sustainability had multiple meanings in the community (a topic to which I give more detailed attention in Chap. 5), but in general terms advocates argued for a balance between conservation goals and community needs, and the greater participation of diverse groups in community decision-making. Admittedly, the protectionist stance has been quite successful in terms of marking out vast territories that are entitled to legal and administrative protection from either NGOs or the state, and although local organisations anecdotally pointed to continuing problems with hunting, poaching and logging in protected areas, this was generally believed to be under control (Wheelwright 2000: 424). Alongside the credit given to local organisations and individual scientists for this success, however, there was also considerable resentment from some community members regarding the relative lack of access to research results and data collected about local ecosystems. Several initiatives were begun by local groups in 2003 to promote wider access to this body of information, including plans to create community information centres, libraries and public consultations, but access to research results remained a point of significant contention.[4] In particular, reports and publications are commonly produced only within academic publications and in the native language of foreign researchers (usually English).

By acknowledging some of the problems surrounding the communication of research results to the wider community, I do not mean to suggest that the local influence of natural science research has been largely negative. In fact, the social and economic benefits of scientific study and conservation have in many ways been overwhelmingly positive. Not only was scientific interest the catalyst behind the region's growth, but visiting and resident scientists have exerted enormous influence on preservation efforts through the establishment and continuing management of

---

[4] I should note here that the issue of communication of research results is not limited to work in the natural sciences. There is also a history of foreign social science researchers arriving in the community to conduct short or long-term research and then failing to send reports or data back to community organisations.

local conservation organisations. These organisations, in turn, have offered protection to huge forested areas, and have been responsible for developing infrastructure to facilitate research in tropical biology and ecology. They have also had an important influence on local decision-making in a region which has historically remained relatively isolated from the state. While some residents complained that this lack of a state presence allowed for too much unregulated development, many national observers also noted a strong correlation between the influence of scientific organisations and researchers and the region's success in conservation and tourism promotion.

NGOs and the Growth of Local Environmental Education

Conservation NGOs have therefore been at the root of Monteverde's growth and development as an international site for tourism, research and conservation since the 1970s. This growth is largely attributable to large-scale land purchase by the Monteverde Reserve, the Monteverde Conservation League and the Santa Elena Reserve. Taken together, this large area of protected forests constitutes the region's largest attraction for scientific researchers and tourists, and lies at the centre of the community's economic well-being. Since the mid-1980s, in addition to land purchase and research, each of these groups have also funded educational programming in an attempt to increase local, national and international support for forest conservation and research.

When we spoke in 2003, Humberto Villa, one of the region's first environmental educators, recalled that the first 'structured' environmental education project in Monteverde was established by a local coffee cooperative, *Cooperativa Santa Elena* (known locally as the *Coope*), in 1985. Humberto joined the effort himself in 1988 when he took a position with the Monteverde Conservation League – the first of the local conservation organisations to devote resources to education. Although he was quick to tell me during our first conversation that much of his knowledge about education has come from working in the field rather than from academic study (he originally trained as an agricultural engineer), Humberto had been involved in local environmental education efforts for more than a decade and was well-placed to outline its historical development.

He explained that after the Conservation League was founded in 1986, the organisation began its work in education by supporting the *Coope*'s existing environmental education project – an annual competition between the region's primary schools. Each year, a theme was established upon which teachers, parents and students focused some of their studies throughout the academic year. Examples included 'the birds of Monteverde and their habitat' and 'the history of Monteverde's forests'. At the end of the year, a jury visited each school to view students' projects and later a 'cultural day of sharing' was held during which prizes (books, sports equipment, gardening tools) were awarded to the schools with the best projects. The *Coope*'s project organisers aimed to provide a focus for environmental education in the schools, and also to teach students how to work together according to the principles of cooperative organisation.

By 1988, the League was already receiving a flood of international funding – including bi-lateral grants and proceeds from a debt-for-nature swap – which were used for land purchase and to significantly expand the environmental education programme to work with schools, youth groups, and local farmers. Within the region's schools, the organisation supplemented the existing national curriculum, used visits and guided tours to encourage students to enjoy and respect local protected areas, and conducted teacher training sessions. These classroom interactions were further supplemented with what Humberto called 'applied environmental education' through which students studied 'daily life' issues such as local agricultural production and its affects on the environment (particularly pesticide use), waste management systems (including application of concepts like '*reusar, reciclar, rechazar*' – roughly translated as 'reduce, reuse, recycle'), and the environmental impacts of high levels of consumption.

The League also sought to connect with young adults outside of school settings by linking its programmes to youth groups organised by local Catholic churches. According to Humberto, each community at that time typically had a youth group which met weekly for socialising and religious discussion. The youth groups were especially popular in isolated communities where there were few opportunities for young people to meet and socialise. Educators from the League used these meetings as an opportunity to discuss local environmental issues and to get young people involved in hands-on work such as the League's projects on waste management and reforestation. When not in the field, educators also led the youth groups in theatre work exercises, helping them to create plays and skits on environmental issues and then taking them to other communities to share. Humberto admitted that it was somewhat 'casual' in terms of formal environmental education programming, but:

> *I did some theatre work myself in university and I thought it was a very strategic way of doing it. People were willing to accept it as a form of recreation, but they were also hearing environmental messages.*

Another piece of the League's work in environmental education focused on local farmers, primarily through the promotion of its reforestation programme. The programme provided seedlings and technical assistance for local farmers to establish windbreaks on their land. These forest fragments took away relatively little productive land, but helped to substantially increase milk and crop yields by sheltering livestock and crops from high winds, acting as a new source of wood for fuel and construction, and providing migratory birds with forest corridors leading to local protected areas. In addition to these benefits to local farmers, the project provided educators from the League with opportunities to engage directly with local landowners and producers. Humberto recalled,

> *In many ways, it was lucky that we [the League] were involved in reforestation projects, because it was through these projects that we were able to reach farmers with environmental messages. On the one hand, we were teaching the farmers a practical skill like how to build a windbreak to protect crops and livestock, but it also gave us a starting point for talking to them about deforestation and conservation in general.*

Beginning in the early 1990s, the Monteverde Reserve and the Santa Elena Reserve also began providing environmental education programmes in the region. When the League experienced severe financial difficulties and discontinued its environmental education programme in 1995, the two reserves took the lead in working with schools and other community groups.

I will return to discussion of those organisations and their activities in more detail in the following sections, but it is useful to note that one of the most interesting features of my conversations with Humberto was his recollection of conflicts over definitions and practices of environmental education theory within the League itself. He recalled that throughout his time with the organisation two fundamentally different types of approaches to environmental education were actively debated. On the one hand were those who argued that environmental education should be exclusively about teaching biology and ecology, while on the other were those who believed that teaching the sciences was useless unless it was paired with an understanding of the daily lives of the people in the communities where programmes are implemented. As Humberto himself argued:

> *In order to educate about environmental issues within a community, you have to know something of the culture, local economy, and legal and political factors that effect it. Otherwise, you can give people all the information you want, but you won't get anywhere in terms of changing behaviours.*

These internal debates reached a turning point in 1991, when there was a serious conflict between educators and members of the League's board – many of whom were scientific researchers. In the end, the two opposing sides reached a compromise of sorts and agreed that the League's programmes should both teach natural history and give attention to local social and economic concerns. This internal conflict is worth noting because it foreshadows identical debates in the community that continued during the time of this research. Specifically, the diversity of perspectives and programmes on environmental education from local conservation groups and individual activists in Monteverde revealed strong tensions between the perceived need to encourage the growth of scientific knowledge and to teach about socio-environmental concerns.

A Tale of Two Reserves

Since the collapse of the League's environmental education programme in the late 1990s, the Monteverde Reserve and the Santa Elena Reserve have led conservation, education and community development efforts in the region. Both organisations officially state that their work encompasses research, conservation and education, but in practice each one has had varying degrees of success in securing funding and other resources for their implementation (cf. Aylward et al. 1996; Wearing and Larsen 1996).

As outlined above, the Monteverde Reserve is a private entity owned by the Tropical Science Centre (TSC), an international research and conservation consortium based in San José. In total, TSC owns a network of reserves which includes the

Monteverde Reserve and three other small properties elsewhere in Costa Rica. The reserve largely maintains its independence from its parent organisation, at least partly due to its distant location. This autonomy extends to its environmental education programme, which is designed and implemented without significant influence from policy makers in the TSC's national office. The rich history of the founding of the Monteverde Reserve has been recounted in great detail elsewhere (cf. Honey 1999; Nadkarni and Wheelwright 2000; Wallace 1992), so I will not devote more space to it here. However, it is relevant to note that when the reserve was first established in the 1972 it was almost exclusively for the purpose of protecting the area's impressive biodiversity and conducting scientific research, with tourism and education as secondary concerns.

The environmental education programme was formally established in 1992, 20 years after the reserve's creation. The programme is funded by proceeds from natural history tours and reserve entry fees, and during the time of this research was administered by a single environmental education co-ordinator, a Costa Rican biologist and educator called Maria Rodriguez, whose responsibilities were large and varied. While the bulk of the programme's work was conducted with local primary schools, for example, Maria was also responsible for hosting visiting groups at the reserve and training local natural history guides. Visiting groups were typically composed of foreign researchers, Costa Rican university students and professors, or tourists. Local school groups were also hosted on-site, but these visits tended to be infrequent because of the associated costs involved, in terms of both money and time. As a result, a significant portion of the environmental education programming was composed of visits to local schools during which students were provided with classroom lessons on environmental topics or support for projects such as school gardens or recycling programmes. These visits were provided upon request – and without charge – for local teachers who asked for help in teaching environmental education or science topics. In 2003, these programmes were mainly conducted for state-funded schools, although there was also some limited co-ordination with local private schools. Additional support for local schools was also offered in the form of teacher training sessions and the use of a mobile environmental education library that included books, games, puppets and other activities. As all of the region's state schools were chronically short of resources, including books, writing materials, and other equipment, local teachers considered these materials to be a significant contribution.

While the Monteverde Reserve's programmes were concentrated on the region's state-funded primary schools, the Santa Elena Reserve's environmental education programmes were focused on projects at the *Colegio* in Santa Elena, to which it is administratively and legally linked. The reserves directed their educational efforts to different schools in the community because local environmental educators, frustrated by overlapping provision in the past, agreed to this division of labour. As a result, in 2003 the majority of the Monteverde Reserve's programmes were conducted in state primary schools, while the Santa Elena Reserve's programmes were focused on the *Colegio* in Santa Elena. In both cases, however, these programmes were perceived by school educators to fill a significant gap of some kind – for instance, of training, resources, or expertise.

In addition to the *Colegio*'s formal environmental education programming (discussed in the previous chapter), students and teachers also received learning and teaching support from an environmental education co-ordinator employed by the Santa Elena Reserve. During the time of this research, the co-ordinator, a Costa Rican educator and conservationist called Luis Delgado, organised seminars and workshops on environmental education topics for both students and teachers, an on-site waste management and recycling initiative, natural history guide training, and plans for the construction of a greenhouse. His responsibilities also required active participation on a variety of inter-institutional and community development commissions, as well as more hands-on work in conservation, maintenance, and planning for the reserve itself.

Throughout my year in the community, I witnessed these two dynamic educators dedicating long hours and enormous energy to their respective organisation's initiatives. Their differing programmes and approaches were suggestive not only of individual differences in perspective, but also of they ways in which they were involved in particular political, economic and social networks within the community. Nonetheless, both Maria and Luis expressed clear ideas about how best to employ environmental education as a tool for bringing change to the community, and each was both supported and constrained in their efforts by the circumstances in which they worked.

Promoting Ecological Knowledge

In many ways, the Monteverde Reserve is an ecotourist's dream. When you arrive at the gates – after the very long and bumpy drive up the unpaved road from Santa Elena – you are an almost immediate witness to the sights which draw so many visitors every year. To begin with there are the green, glossy, misty and overpoweringly dense cloud forest growth overarching the visitors' centre, frequent visits by hummingbirds and coatis to the covered picnic tables nearby, occasional glimpses of howler monkeys and sloths in the trees overhead, and usually a local nature guide armed with a telescope to show the curious visitor other, less ostentatious, natural treasures. Visitors are drawn to the reserve by stories of sightings of all of these creatures, as well as of rarer ones like the Resplendent Quetzal, whose protection was another of the early catalysts for the reserve's establishment. It is arguably a perfect site for visiting nature lovers, and also for teaching young people and visitors to admire and respect the cloud forest. In addition to the extensive network of nature trails which crisscross the property and provide views of both sides of the Continental Divide, the reserve's facilities include classrooms, a lecture hall, a hummingbird garden, and a laboratory located just inside the edge of the forest. In 2003, the reserve counted on a staff of around 30 employees, including a director, administrative and communications staff, and protection and maintenance staff, the majority of whom were Spanish-speaking Costa Ricans.

For the first few weeks of my stay in the community, I had arranged to stay in one of a few rooms on-site which the reserve provides for visiting researchers. After briefly settling in, I met the reserve's director to discuss my research plans and he kindly arranged for me to meet the environmental education co-ordinator, Maria Rodriguez, that same afternoon. She immediately struck me as an impressively energetic and passionate educator. Originally from San José, Maria was at the time only in her mid-thirties but had already been living and working in Monteverde for more than 10 years. She first trained as a biologist, and then later received training in environmental education both in Costa Rica and the US. She had held her position at the Monteverde Reserve since she arrived in the region, sometimes with additional assistance from other reserve staff, but most often working alone. Her long-term residence in the region made her extremely knowledgeable about the area's ecosystems, endemic species, and environmental problems, and it also brought her into close and sustained contact with students, educators, researchers, and staff from other community organisations. She believed that this familiarity had proven useful both in terms of the impacts of her programmes on student learning, as well as her ability to collaborate effectively with fellow educators. Local primary school teachers, in particular, relied heavily on her for curricular support and special programming. In many cases, arrangements for student programmes were initiated by a phone call or personal request from a local teacher, and were tailored to their specific teaching needs.

On one occasion, for example, a fourth grade teacher from *Escuela Santa Elena* phoned to arrange a time for her students to visit the reserve and learn about a few concepts in the science curriculum: habitat, niche, community, ecosystem, equilibrium, mutualism, communalism and parasitism. During the few days prior to their arrival, Maria created and assembled games, worksheets and activities designed to teach these concepts in an engaging way.

On the day of this visit, a group of about 25 children arrived at the Monteverde Reserve's entrance and piled noisily out of the back of one of the reserve's trucks (this had been provided because the school was not able to provide its own transportation). The students all wore the required state school uniform – navy trousers and white tops with dark shoes – and were animatedly excited about their visit. Once some order could be restored, we walked as a group along a forested trail to an open area where there was enough room to do our first activity. For this exercise, the students were put into groups – 'monkeys' (*monos*) and 'habitat' (*habitat*) – and instructed to line up in two rows facing away from each other. Members of the habitat group were instructed to choose to be either water (*agua*), food (*comida*) or shelter (*refugio*), and shown hand signals for each choice. At the count of three, the groups were to turn to face one another, then the 'monkeys' should run as rapidly as possible across the field to claim a partner in the habitat group as their 'resource'. These captured resources then became members of the monkey group. As we played the game over several rounds, the students began to notice that there was less and less 'habitat' to go around. Monkeys that were unable to find a food, water or shelter were instructed to fall down and 'die' and then become part of the 'habitat' group. In the last round, Maria gathered the 'habitat' group and quietly asked them to fall to the ground at the count of three rather than running to meet the other group. She told them this story: *'There once was a man who owned some land. One day he decided to cut the trees so there was no longer food, water or shelter for the animals that lived there.'* When the children acted out this scenario, the other group came running over, shouting questions. We sat the children down in a circle to discuss what they thought had happened. They offered their own opinions and then listened carefully as Maria explained how changes in the population of a species are linked to changes in habitat – especially through drastic change caused by human intervention.

Our next activity took us up to the reserve's classroom – essentially a small wooden cabin in the forest – which contained tables, chairs and a blackboard. Along two walls were windows looking out onto the surrounding forest, while the remaining surfaces were covered by a painted mural depicting a forest scene filled with plants and animals found in the reserve. In addition, there were also informational displays about soil erosion and Quetzal preservation, and a large model relief of the Monteverde region. After we arrived, Maria sat the children down in a circle and asked them to place fabric animal cut-outs into the appropriate parts of a fabric-covered board 'habitat'. The children were so eager to participate that she had to repeatedly ask them to sit down so that everyone would have a chance to see. After the animals were all correctly placed, she began a discussion of the different kinds of 'jobs' – omnivore (*omnivoro*), carnivore (*carnivoro*) or herbivore (*herbivero*) – each of these animals does and likened this to the different jobs that each of their parents does in their own, human community.

By now the children were beginning to become restless, so we took a short break before starting a new activity that centred on the concepts 'animate' and 'inanimate'. Maria asked the students to join her again in a circle on the floor and then used a handmade set of flashcards to explain the difference between the two concepts. Each card contained the name of either a plant, animal, or inanimate object (e.g. water, air, soil). If the card showed something animate the students should stand, and if not they should crouch down on the floor. Those who answered incorrectly were 'out'. The group jumped up and down happily, playing the game until only one student was left. He then went through the entire stack of about 20 cards alone and got them all correct again, to the delight of his fellow students.

Partly as a way of calming down the now quite excited group of youngsters, Maria moved quickly into the final activity. Earlier that morning, before the group arrived, Maria and I had marked out three miniature 'habitats' with plastic tape – around a medium-sized plant, and underneath a large stone and a fallen tree trunk – in the forested area just outside the classroom door. The class was split into three groups and each was given a sheet of paper on which they were to record what they found in and around each area. This was intended, Maria had told me earlier, as an exercise in careful observation. The teacher, Maria and myself each led a group and guided their observations at the three sites – asking questions like 'what do you see?', 'what is different here than in the other two areas?', 'what is similar?' – and then we gathered together again in the classroom to discuss their findings. While many of the group talked enthusiastically, the students had by this point been at the reserve for nearly half of the day, and some students seemed to be losing interest. Although they all participated in the exercise, for some it was with significantly less enthusiasm than in the earlier activities. Maria finally called an end to the activities around lunchtime, and we said goodbye to the students and their teacher as they climbed into the truck for the journey back to their school.

Educational Approach and Impacts

In this half-day of programming, as in others I took part in during the year both at the Monteverde Reserve and other local sites, it was sometimes difficult to maintain student interest and order. Of course, student discipline and effective teaching methods are central topics of conversation for every educator, and do not necessarily compose a particular set of problems for environmental education. In many cases, of course, interest and discipline also varies greatly depending on the group. As a case in point, later that same day at the reserve we conducted these activities again with the same teacher's cohort of afternoon students, and they were highly interested and well-behaved throughout the session. Regardless of these challenges, however, I participated in many exchanges that suggested that Maria's environmental education programmes had significant impacts on student learning.[5] For instance, during the many school visits, meetings and other activities on which I accompanied Maria throughout the year, I regularly observed school children and young people enthusiastically greeting her and asking when she would next visit their school. They often remembered the games and activities that they had done with her previously and pleaded for more lessons or opportunities to visit the reserve to walk in the forest. On occasion, she would quiz them about the environmental topics they had discussed or ask them if they were recycling in their homes and schools. More often than not, students would respond with confident answers and an apparently strong understanding of the issues they had studied.[6]

Maria and I had frequent conversations over the course of the year about how best to engage children and young people in environmental learning. In her daily practice, Maria relied on a limited number of resources in the reserve's environmental education library (which she had compiled herself over the course of her time there) which contained a mixture of Spanish and English language publications created by the Costa Rican Ministry of Education and the national universities, as well as international organisations such as UNESCO, UNEP, and IUCN. She also explained that

---

[5] Whether these impacts are strictly measurable is an separate issue and one which has received significant attention both from educators in Monteverde and within the academic literature (cf. Rovira 2000; Fien et al. 2001).

[6] Whether this knowledge also changes behaviours – by encouraging recycling or stopping illegal hunting or logging – is similarly difficult to measure and has been the subject of intense debate in the field (cf. Kollmuss and Agyeman 2002; Gough 2002; Courtenay-Hall and Rogers 2002).

her programmes were heavily rooted in training she had received during a 2 month-long workshop in 1997 with Professor Sam Ham at the University of Idaho (USA). Professor Ham's work on 'environmental interpretation' (cf. Ham 1992; Ham and Weiler 2002) is internationally-known, and frequently praised, by educators, protected area managers, and conservationists. It is a style of presentation designed specifically for educators working within nature reserves, and with natural resources (trails, observation sites, etc.) close at hand. According to Ham: 'Environmental interpretation involves translating the technical language of a natural science or a related field into terms and ideas that people who aren't scientists can readily understand' (1992: 3). This style of education requires educators to select relevant facts and concepts which can then be used to pass on a specific idea: 'In interpretation... the goal is to communicate a message – a message that answers the question "so what?" with regard to the factual information we've chosen to present. In this respect, there's always a "moral" to an interpreter's story' (Ham 1992: 4).

When viewed in these terms, the overarching 'moral' of Maria's programmes at the Monteverde Reserve was that rare and pristine natural areas – such as the cloud forests within the reserve's borders – must be protected from human intervention (synonymous in this case with destruction) for the good of humankind and the advancement of scientific knowledge. The central underlying goal of these programmes, therefore, was the conversion of students into 'environmentally responsible' citizens and community members. This perspective is somewhat reminiscent of a approach to environmental education, most frequently employed by conservation NGOs, that has already received significant attention from research:

> 'Education as perceived by most within the environmental movement would, if possible.... [be] a process of initiation where people would see the world as the environmentalist or conservationist would have them see it. And, as well as the central importance of introducing people to first-hand experiences, such education would also be augmented by a more fundamental, rational approach to the environment, using, in the main, ecology and the natural sciences to help demonstrate the scientific validity of the arguments for better environmental management.' (Martin 1996: 44)

What it is perhaps most important to note about this approach to environmental education is the emphasis it places on scientific ways of knowing, understanding, and protecting the environment. It tends to neglect, on the other hand, other ways of knowing or relating to the natural world, as well as the many social, cultural and economic factors which influence human behaviour and relationships to it (cf. Wals 1996; Sterling 2001). Furthermore, the prime place given to scientific knowledge within the reserve's educational programmes was clearly linked to both the organisation's long-term efforts in the strict protection of its forested lands, and its own social and economic location within the community.

The success of regional conservation programmes by the Monteverde Reserve and other local organisation – in terms of biodiversity protection, research, and tourism – during the last 30 years has, however, brought with it some unexpected effects that have required a reassessment of local management styles. The rapid growth of the local population and steady increases in tourism visitation, for instance, have forced local conservation groups to shift their focus from a single-minded concern with strict protection of forested areas to a much greater attention to the impacts of local development issues such as water and waste management.

This reassessment has also been reflected within many organisations' environmental education programming. As Maria recalled:

> At first, all of my environmental education programmes focussed almost totally on biological concerns, and especially on the need to protect forests from human activity, as a support for the reserve's conservation goals. But in the last few years, because of the growth of the region's population and infrastructure, we've had to start giving greater attention to issues like the region's insufficient water and waste management systems, and how they impact the reserve's forests and the success of the local tourism industry.

As part of these efforts, in 1996 the Monteverde Reserve established a pilot recycling programme which accepts glass, paper, aluminium, and plastic products. Maria had taken on much of the responsibility for promotion of the programme, and during 2003 she was actively organising lectures and activities about it for local schools and other groups. In one workshop that I attended, she gave a short lecture on existing recycling programmes in Costa Rica as a whole and an introduction to the Monteverde Reserve's project in particular, and then presented the student participants with a pile of materials, asking them to first sort out which of them were recyclable and then organise them by type. As with Maria's other, more specifically biologically-focused environmental education work, much of the emphasis of these events was on raising awareness of the human causes of environmental damage, and responsible ways to ameliorate it.

The Monteverde Reserve is only one of many local organisations whose efforts continue to be dominated by such scientific styles of discussion, management, and education, yet there are also opposing voices within the community. One local environmental educator, for example, argued strongly that merely teaching about the natural sciences was too limited an approach for environmental education programmes, and that attention must also be given to the social and economic factors that effect local environmental management:

> 'Environmental educators must facilitate processes that promote the capacity of the community to select habits, develop forms of sustainable production, and resist the objects that the promoters of the consumer system make us believe are necessary. In the long term, the protection of natural ecosystems and their biodiversity will be favored to the extent that the local population develops attitudes, knowledge, and skills to sustain development in Monteverde. The community will develop its potential to form a concept of life that is fuller and more sustainable to the extent that environmental education is a conscious process of the people and reaches beyond programs of conservation institutions.' (Vargas 2000: 378)

This perspective on environmental education echoes the work of many international researchers and educators who have sought to promote programmes which more strongly connect environmental concerns with social realities (cf. Huckle and Sterling 1996; Bourn 2008). This work is also indicative of the broader emphasis on sustainability within discussions of international development since the mid 1980s (cf. Redclift 1987; Warburton 1998).

Training the Next Generation of Educators

Luis Delgado, environmental education co-ordinator for the Santa Elena Reserve, actively endorsed this more socially-oriented view of environmental education. Like many of his colleagues in Monteverde, Luis' experience in conservation and education was extensive, including work for the Ministry of Environment and Energy in several national parks, as well as with various other private conservation organisations around the country. Although he was not born in Monteverde, he had spent many years living and working in the region, including in the Monteverde Conservation League's environmental education programme when it was at the height of its influence. In one way or another, he had been actively involved in education and conservation in various parts of Costa Rica for the previous two decades.

I first met Luis in February 2003, when we were introduced by a colleague from the *Colegio*. His work was based at the Santa Elena Reserve's main administrative office and tourist information centre – a small building situated along the main road and on one corner of the *Colegio*'s grounds. This small building contained offices for the reserve's director, Luis, and two staff in reception, while the majority of the organisation's employees (park guards, maintenance staff, administrators) were based at the reserve itself. This location placed the *Colegio* office some distance from the reserve, but within easy walking distance of the centre of the village of Santa Elena. As a result, the office was frequently busy with visitors, including students, teachers, tourists and staff from other local organisations.

Luis' position at the Santa Elena Reserve involved both work with the *Colegio*'s students and teachers, as well as co-ordination of meetings and projects with other local organisations. As part of his work within the *Colegio* itself, Luis led workshops and seminars, and guided students on visits to the Santa Elena Reserve either for hikes on the trails or to have them work as volunteers. Teachers at the school – and especially in the ecotourism specialisation (described in Chap. 3) – also relied on him to assist with the development of lessons on local ecology and endemic species, and would regularly request information, ideas and workshops for their students. In addition to this support for classroom learning, he worked to implement a set of what he called 'applied' environmental education projects, including plans to build a greenhouse and to design a sustainable system to manage waste from the *Colegio*'s agricultural and livestock projects (including pigs, cattle, chickens and a fish farm).

Throughout our many conversations during the year, Luis expressed a strong belief that environmental education is about more than simply teaching information:

> *When I started working in environmental education fifteen years ago, most programmes focused on teaching information about environmental issues, but after a few years I started to wonder if this was enough to achieve change. Now I believe that discussion of environmental topics has to be connected to the social reality in which people live.*

Underlying his efforts was also a firm conviction that the central purpose of environmental education is to encourage people to take action: 'Environmental education is a process, not a well-defined thing. It is a road we are making as we go along.

The three main steps in this process are knowledge, consciousness raising and action' ('*conocimiento, conscientización y acción*'). A significant portion of his work at the *Colegio*, therefore, was aimed at raising the students' consciousness about environmental issues and encouraging their active engagement with social issues. Through workshops and lessons with students in the ecotourism specialisation, he told me, he sought to 'educate the next generation of environmental educators'. Interestingly, he interpreted the term 'environmental educator' rather broadly, and included both those who would work directly as educators and those who would be employed for conservation organisations and thus involved in public education initiatives more generally.

About a month after our first meeting, Luis invited me to attend a series of workshops he had organised for three classes of students in the ecotourism specialisation (tenth and eleventh grades). During that day, I observed and assisted him as he conducted two-hour sessions with each group in their classrooms. The goal of the day, he told me, was to teach the students more about the realities they would face working as environmental educators in the future. Specifically, he emphasised to the students the importance of effective communication and the need for an understanding of the different community groups with which educators work. He also engaged the groups in a pragmatic discussion about the potential for educators to bring about social change. In the main, Luis led these sessions by himself, although the teacher responsible for each class occasionally added comments regarding a particular topic or the ways in which the discussion was linked to previous classroom lessons. The students were in many ways typical of young adolescents everywhere – independent, strong-minded, and occasionally mischievous, but sometimes also uncertain and preoccupied with music, fashion, and school gossip. They also displayed a sincere interest in their studies of ecotourism and conservation, and greeted Luis enthusiastically as he entered their classrooms.

He began each workshop by saying,

> 'Instead of me giving information through lectures, we'll learn together by doing a set of activities…. Environmental education isn't about giving lectures, it's about action…. The two steps in environmental education are: 1. to inform, and 2. to take action. Environmental education should be dynamic and active and constantly changing, and the most important part is taking action for change'.

The first activity he organised was an exercise in communication. To begin, we placed the students in pairs and asked them to sit face-to-face and talk normally about any subject. After a few minutes had elapsed, Luis began going around the class, asking one person in each pair to turn their back on the other, and then to continue talking as before. After a few more minutes he stopped them and asked what they thought of the experience. Several of the students said that it was much harder to speak without facing one another because they were not able to see the face of the person they were talking to. In response to this, Luis pulled several small hand-held mirrors out of his bag and asked them to try using them so that while one person's back was still towards his or her partner, by holding the mirrors, they would be able to see one another's faces. The pairs tried this for a few minutes as we circulated around the room and when Luis asked for reactions this time, they commented that it was still difficult to communicate comfortably. He then asked them to return to speaking face-to-face as they had originally, but with one student from each pair standing on his or her chair. He allowed them to speak this way for a few moments, and then asked them to switch places. Afterwards we gathered the students together as a large group and Luis asked them to reflect on how this experiment went. More than one of the students responded that it was difficult and uncomfortable, and as they talked it over among themselves, Luis opened up the discussion further to talk about unequal power relationships in the office, in families, and in other situations in life, and commented that as educators they would have to be aware of these differences and be sensitive to them.

After finishing this and a few other exercises on communication, Luis began a new activity to illustrate the challenges of decision-making. To do this, we put the students in groups and asked them to consider a hypothetical situation: *Imagine yourselves in a boat in the middle of the ocean. The only way for most of the group to survive this ordeal is if one member of the group is thrown overboard. Who should go over?* The groups were given 10 min to discuss the situation and agree on some criteria for making this difficult decision. Of the groups that I observed, the conversation revolved around how to make the most 'fair' decision, and/or the relative importance of certain group members to the survival of the group as a whole: the strongest, the smallest, or those with special skills.[7] After the allotted time elapsed, Luis asked each group to explain to the rest of the class how they had made a decision. Many groups had decided by lottery, and in one group a member had even agreed to go over for the good of the rest of the group.

After each group had shared their thoughts, Luis asked those who had been 'thrown overboard' to talk about how they felt. Some were unhappy about being chosen for what they saw as 'unjust' reasons, but those who were chosen by lottery responded that it was simply bad luck. Luis asked them:

> But is this realistic or should we make difficult decisions based on more relevant criteria?... This is something that you will all have to face as environmental educators and human beings – and especially because as young people you will be faced with lots of difficult decisions. When you are doing environmental education you have to think carefully about how best to help a community... You cannot help everyone, and there are always going to be people who don't like your work or who you cannot reach. You have to consider all the relevant factors and not allow your own prejudices of race or gender or anything else to colour your judgement.

As the day continued, we conducted these exercises with another tenth grade class and then finished the day with a class of eleventh grade students. In this last session Luis additionally asked them to create skits of about 3–5 min in length.

---

[7] On one occasion, for instance, one young woman was saved from being thrown over because – as she was told by the rest of her all male group – she could 'cook for the rest of us'. I have not dealt extensively with gender issues within the schools as part of this research, although they undoubtedly have a significant impact on environmental education and management. A significant body of literature on gender issues in Costa Rica does exist which explores related issues (cf. Stocker 2005; Twombly 1998; Leitinger 1997b; Palmer and Chaves 1998).

They were given 10 min to work on a skit about an assigned theme, and then we gathered them together as a large group to see what they had created. The first group's theme was 'Contamination in Monteverde' and they chose to portray it by pretending to be what they identified as a 'typical' tour group visiting Monteverde. They pulled together some chairs and assembled themselves as if they were riding on a bus with a 'guide'. The 'tourists' on the bus made negative comments about rubbish on the side of the road or in roadside streams, and asked the guide to explain why it was there. The guide brushed their questions aside, and instead pointed to other items of interest such as birds, trees and butterflies, but the tourists continued to insist that he answer their questions. At one point, one of the 'tourists' pointedly spoke to the guide in English and asked him to explain something. The rest of the class laughed loudly at this locally familiar image of a demanding American tourist who does not speak Spanish.

The second group dealt with the theme 'Say "No" to War'. This they did in silence by enacting a scene in which two of the young men began to have a friendly conversation that eventually turned into a disagreement. One of them walked away in anger, then came back carrying a 'gun' and started shooting his 'enemy' as well as the others in the group. One of the young woman was finally able to reach him and convince him to put the gun down. She 'spoke' to him about what he had done, and he realised the error of his ways and carried his fallen enemy away to get help. The last group had created a skit around the theme 'Young People and Conservation'. They acted out a scene in which a group of students was planting trees in a reserve. As they were working they heard gunshots and found a young woman 'hunting' nearby. They convinced this young woman that hunting was wrong, explaining that the animals should be left in peace within the reserve's boundaries. In the end, she joined them in planting trees.[8] After each skit, there was discussion amongst the whole group about the themes and their meaning to each student individually and to the community. When the activities were all finished, we said good-bye to the students and their teacher, and they politely thanked Luis for organising the session with an enthusiastic round of applause.

Educational Approach and Impacts

Throughout the year, as is evident from the description above, in addition to teaching or conducting projects about specific environmental issues, Luis spent a significant amount of time and energy teaching students about the connections between environmental management and social concerns. He stimulated discussions, for example, on the impacts of unequal power relationships and social and economic difference on environmental management decisions as well as individual behaviours. He also

---

[8] Each of these skits reflected the key discourses of peace, non-violence and conservation which I frequently heard emphasised both within the community and nationally. The piece on war was particularly striking in the context of on-going US military action in Iraq at the time. There was significant disapproval of this within the Costa Rican media as well as in everyday conversation.

encouraged students to develop a personal awareness of these issues by arranging exercises and activities in which their impacts could be experienced in classroom settings. When conducting the boat scenario activity, for instance, he deliberately organised groups with unequal gender representation (one young woman in a group of young men or vice versa) or asked a teacher to join a group of students.

Luis' perspective and approach draws on the work of transformative educationalists both within environmental education and in education more generally. In Latin America, this particularly includes educators such as Paolo Freire whose work is strongly linked to the wider tradition of liberation theology and to movements against economic, ethnic, and gender oppression (cf. Freire 1972). Perhaps Freire's most long-lasting contributions are his insistence on the revolutionary character of education, and his critique of what he labelled 'banking education' – wherein students are treated as empty vessels waiting to be 'filled' with information. Teachers in this relationship 'hold' knowledge and present it to students as an unchanging and unchangeable unit, often without allowing space for questioning or criticism. Freire argued that this approach serves to reinforce the status quo of the dominating groups from which such pedagogical structures and practices originate. Educational programmes which allow space for self-directed learning, questioning and critique, on the other hand, provide opportunities for students to deconstruct power structures and to take action to change them.

The work of Freire and others has also been influential in environmental education research internationally, as illustrated, for example, in socially-critical styles of environmental education which argue that: 'The goal of education is the optimal development of people, with an emphasis on autonomy and critical thinking' (Sauvé 1996: 1). Advocates of transformative approaches also argue that the most important role of education is to raise awareness and change attitudes about issues of concern to humanity (cf. Wals 2007; Sterling 2001; Huckle and Sterling 1996).

In Monteverde, such a perspective on environmental learning stands in real contrast to programmes (like those of the Monteverde Reserve) which focus on the provision of particular information about local ecology and the promotion of a pre-determined environmental ethic and 'good' behaviours. It does, however, fit well with the overall agenda of the Santa Elena Reserve – with its focus on locally-controlled community development – and with the organisation's own social and economic location within the community.

Diverse Approaches, Diverse Positions

The contrast between these two approaches to environmental learning is clearly striking. As noted in each section, these differences are not simply related to the educator's individual perspectives, but also reflect the wider agendas and positioning of the NGOs for which they worked.

The Monteverde Reserve, for instance, is a relatively large organisation with a strong financial position, and it receives broad support from national and international scientific organisations and researchers with an interest in the preservation of biologically valuable areas. It also counts on a more than 30 year history of successful international tourism promotion in the region. Scientists and conservationists within the organisation are able to call upon influential networks of international colleagues, including contacts in foreign universities and research institutions, as well as international conservation or policy organisations, in order to gain access to funding or support. They can also mobilise these networks in order to impose pressure on local or national policy makers, or – as has happened in the past – to put a stop to local development projects of which they did not approve.

The much smaller Santa Elena Reserve, on the other hand, was established relatively recently and has yet to receive substantial attention from either the international or national scientific community, or the international tourism industry. It does, however, count on significant support from regional and national policy makers as well as national tourism business groups. Such affiliations provide both financial and institutional support for local projects, including educational programmes, and provide space for the active contestation of the historically-dominant protectionist environmental management agendas of other groups in the community.

These differences – both between the two educators' individual perspectives and the wider agendas of the two NGOs for which they worked – are also indicative of an even more significant rift in the community regarding local development and environmental protection. Specifically, while both Maria's and Luis' efforts were widely appreciated by local school teachers – who were grateful for support in filling existing (and on-going) gaps in resources, training and expertise in their schools and classrooms – other sectors of the community had more complicated reactions to, and relationships with, their work. Maria's programmes (and the Monteverde Reserve) generally received support from the local scientific community – a powerful and largely foreign (US) group; Luis' programmes (and the Santa Elena Reserve) were more often supported by local development organisations – a much smaller and largely Costa Rican group.

Historically, local scientists have dominated local decision-making about environmental protection and local development, and their perspectives on these issues have tended to take precedence as a result. Some of the strongest local supporters of the Monteverde Reserve's educational approach have thus often been the most stridently critical of more socially-critical perspectives. For example, the use of theatre exercises or group discussions about social issues – by Luis while at the Santa Elena Reserve and by environmental educators at the Monteverde Conservation League during the 1980s – have received particular criticism from local scientific researchers. As one long-term resident, a biologist originally from the US and who was influential in the creation of the Monteverde Reserve, told me frankly: 'that's just not environmental education'. He continued:

'In the old days the Conservation League had pretty serious environmental... *so-called* environmental education programmes... They tended in my opinion to be pretty leaky in the sense that there wasn't something that sort of clearly circumscribed what we would call environmental education... At least the way I conceive of it, it's a much, much broader and maybe sort of more of a biological or ecological education.'

Local scientists such as the one quoted above have in the past exercised significant influence over local decision-making in Monteverde through their affiliations with external funders and national government agencies, as well as their discursive influence in international publications and other fora concerned with conservation and research. As a result, more socially-oriented styles of education and associated knowledges have tended to be marginalized within the community. The noticeably greater level of support for educational approaches like Maria's from such influential actors should be seen, therefore, as a part of much larger struggles over environmental learning and its links to decision-making about the local management of the environment and community development.

Environmental Knowledge and Community Relationships

As the discussion in this chapter highlights, environmental education programmes organised by local conservation groups in Monteverde were strongly linked to wider social and economic relationships, and to moments of both conflict and collaboration in the community as a whole. In a sense, these negotiations over the style and content of environmental education can be seen as battles over knowledge – of different kinds, for different audiences, and to meet diverse aims.

As I will discuss in the next chapter, diverse community members even more visibly enacted this struggle over environmental knowledge through discussions and debates in public spaces. In these arenas, a range of different groups argued for their legitimate role in local leadership and decision-making by drawing on particular understandings of community development and environmental management, and – in doing so – both directly and indirectly promoted particular approaches to environmental education.

References

Aylward, B., Allen, K., Echeverría, J., & Tosi, J. (1996). Sustainable ecotourism in Costa Rica: The Monteverde Cloud Forest Reserve. *Biodiversity and Conservation, 5*, 315–343.

Blum, N. (2009). Teaching science or cultivating values? Conservation NGOs and environmental education in Costa Rica. *Environmental Education Research, 15*(6), 715–729.

Bourn, D. (2008). Education for sustainable development in the UK: Making the connections between the environment and development agendas. *Theory and Research in Education, 6*(2), 193–206.

Campbell, L. (2002). Conservation narratives in Costa Rica: Conflict and co-existence. *Development and Change, 33*, 29–56.

Courtenay-Hall, P., & Rogers, L. (2002). Gaps in mind: Problems in environmental knowledge-behaviour modelling research. *Environmental Education Research, 8*(3), 284–297.

Fien, J., Scott, W., & Tilbury, D. (2001). Education and conservation: Lessons from an evaluation. *Environmental Education Research, 7*(4), 379–395.

Freire, P. (1972). *Pedagogy of the oppressed*. Middlesex: Penguin Books Ltd.

References

Gough, S. (2002). Whose gap? Whose mind? Plural rationalities and disappearing academics. *Environmental Education Research, 8*(3), 273–282.

Griffith, K., Peck, D. C., & Stuckey, J. (2000). Agriculture in Monteverde: Moving toward sustainability. In N. Nadkarni & N. Wheelwright (Eds.), *Monteverde: Ecology and conservation of a tropical cloud forest* (pp. 389–417). Oxford: Oxford University Press.

Ham, S. (1992). *Environmental interpretation: A practical guide for people with big ideas and small budgets.* Golden: North American Press.

Ham, S., & Weiler, B. (2002). Interpretation as a centrepiece in sustainable wildlife tourism. In R. Harris, T. Griffin, & P. Williams (Eds.), *Sustainable tourism: A global perspective.* London: Butterworth-Heinneman.

Honey, M. (1999). *Ecotourism and sustainable development: Who owns paradise?* Washington, DC: Island Press.

Huckle, J., & Sterling, S. (Eds.). (1996). *Education for sustainability.* London: Earthscan Publications.

Kollmuss, A., & Agyeman, J. (2002). Mind the gap: Why do people act environmentally and what are the barriers to pro-environmental behaviour? *Environmental Education Research, 8*(3), 239–260.

Leitinger, I. A. (1997a). Long term survival of a Costa Rican women's crafts cooperative: Approaches to problems of rapid growth at CASEM in the Santa Elena-Monteverde Region. In I. A. Leitinger (Ed.), *The Costa Rican women's movement: A reader.* Pittsburgh: University of Pittsburgh Press.

Leitinger, I. A. (Ed.). (1997b). *The Costa Rican women's movement: A reader.* Pittsburgh, PA: University of Pittsburgh Press.

Martin, P. (1996). A WWF view of education and the role of NGOs. In J. Huckle & S. Sterling (Eds.), *Education for sustainability* (pp. 40–51). London: Earthscan Publications.

Monteverde Friends Meeting. (2001). *Monteverde jubilee family album.* Monteverde, Costa Rica: Monteverde Friends Meeting.

Nadkarni, N., & Wheelwright, N. (Eds.). (2000). *Monteverde: Ecology and conservation of a tropical cloud forest.* Oxford: Oxford University Press.

Palmer, S., & Chaves, G. R. (1998). Educating señorita: Teacher training, social mobility and the birth of Costa Rican feminism, 1885–1925. *Hispanic American Historical Review, 78*(1), 45–82.

Redclift, M. (1987). *Sustainable development: Exploring the contradictions.* London: Routledge.

Rovira, M. (2000). Evaluating environmental education programmes: Some issues and problems. *Environmental Education Research, 6*(2), 143–155.

Sauvé, L. (1996). Environmental education and sustainable development: A further appraisal. *Canadian Journal of Environmental Education, 1*, 7–33.

Sterling, S. (2001). *Sustainable education: Re-visioning learning and change* (Schumacher Briefing Number 6). Totnes: Green Books Ltd.

Stocker, K. (2005). *'I won't stay Indian, I'll keep studying': Race, place, and discrimination in a Costa Rican high school.* Boulder: University Press of Colorado.

Tosi, J.A. (n.d.). *Una historia breve de la Reserva Bosque Nuboso de Monteverde del Centro Científico Tropical.* Unpublished lecture delivered at the 20th anniversary of the founding of the Monteverde Reserve (1992).

Trostle, J. (1990). *The origins of the Monteverde Institute.* Unpublished internal document. Monteverde, Costa Rica: Monteverde Institute.

Twombly, S. B. (1998). Women academic leaders in a Latin American university: Reconciling the paradoxes of professional lives. *Higher Education, 35*, 367–397.

Vargas, G. (2000). The community process of environmental education. In N. Nadkarni & N. Wheelwright (Eds.), *Monteverde: Ecology and conservation of a tropical cloud forest* (pp. 377–378). Oxford: Oxford University Press.

Vivanco, L. (2006). *Green encounters: Shaping and contesting environmentalism in rural Costa Rica.* New York: Berghan Books.

Wallace, D. R. (1992). *The Quetzal and the Macaw: The story of Costa Rica's national parks*. San Francisco: Sierra Club Books.

Wals, A. (1996). Back-alley sustainability and the role of environmental education. *Local Environment, 1*(3), 299–316.

Wals, A. (Ed.). (2007). *Social learning towards a sustainable world: Principles, perspectives and praxis*. Wageningen: Wageningen Academic Publishers.

Warburton, D. (1998). *Community and sustainable development: Participation in the future*. London: Earthscan.

Wearing, S. (1993). Ecotourism: The Santa Elena rainforest project. *Environmentalist, 13*(2), 125–135.

Wearing, S., & Larsen, L. (1996). Assessing and managing the sociocultural impacts of ecotourism: Revisiting the Santa Elena rainforest project. *Environmentalist, 16*, 117–133.

Wheelwright, N. (2000). Conservation biology. In N. Nadkarni & N. Wheelwright (Eds.), *Monteverde: Ecology and conservation of a tropical cloud forest* (pp. 419–456). Oxford: Oxford University Press.

Chapter 5
Environmental Knowledge in Public Spaces

Abstract Research on environmental education has given extensive attention to diverse theories and practices of environmental education, but the field has tended to take a narrow focus on specific curricula and policies or on activities within strictly defined sites such as schools, classrooms or protected areas. While this has provided useful detail about individual initiatives and the roles of key actors (especially classroom teachers and other educators), often only scant attention has been given to how these are connected to the broader social, economic and political relationships in which they are situated. In contrast, this research set out to understand these relationships and their impacts on perspectives and practices of environmental education. This chapter looks specifically at the ways in which community members in Monteverde, Costa Rica meet, interact and learn about environmental management and community development in public spaces such as meetings and ecotourism destinations. It describes how this knowledge and learning, in turn, influences public opinion, impacts upon the implementation of community projects, and feeds back into ideas about the 'appropriate' content and pedagogical orientations of environmental education in local schools and conservation organisations.

Keywords Community development • Environmental education • Informal learning • Public spaces • Sustainable development

In addition to the more formalised programmes in schools and local conservation organisations described in previous chapters, public spaces also proved to be important sites for exploring perspectives and practices of environmental education in Monteverde in 2003. Locally, these public spaces were of two basic types. The first of these were the many educationally- or environmentally-oriented tourism destinations operating outside of the three main protected areas. Although they are privately owned and operated, these spaces were designed to be sites for public engagement with knowledge about the local environment, and are open to both local residents and visitors. A second key type of space for public engagement were gatherings such as meetings, workshops, seminars, lectures and public consultations where local residents encountered and discussed environmental knowledge and understandings of various kinds.

All of these locations were important to environmental learning, as well as social and community life, in a number of ways. Firstly, they acted as sites for public engagement with environmental learning by offering residents opportunities to learn about environmental topics or other issues of local concern. However, I suggest that these sites not only provided spaces for individual learning, but were also strategically used by many community members in order to speak to local concerns occurring in the community *outside* these sites, and particularly their worries about the character and progress of local development and environmental management.

This argument meshes well with a large body of educational, anthropological and sociological research which shows that – rather than being a simple process of transmission and assimilation by an individual – learning is an inherently social process:

> 'Far from social learning being a questionable appendix to individual learning, individual learning itself is a suspect phenomenon…. as some would argue, there is in reality no individual learning to speak of. Virtually anything one learns, according to the socio-cultural view, comes deeply embedded in a cultural context, involves culturally informed and laden tools, and figures as part of a range of highly social activity systems, however alone the learner may be at particular moments.' (Salomon and Perkins 1998: 16)

Although such a conceptualisation of learning is not without its critics (cf. Anderson et al. 1996) and despite the persistence of 'common sense' understandings of learning as simple knowledge transfer (cf. Hager and Hodkinson 2009; Illeris 2007), there is compelling evidence from research that underscores the importance of the social aspects of learning.

Perhaps one of the most influential publications in this area is Lave and Wenger's (1991) *Situated Learning: Legitimate Peripheral Participation*, which explores the relationships between learning and participation in communities of practice:

> 'Learning is a process that takes place in a participation framework, not in an individual mind. This means, among other things, that it is mediated by the differences of perspective among the coparticipants. It is the community, or at least those participating in the learning context, who "learn" under this definition. Learning is, as it were, distributed among coparticipants, not a one-person act.' (Hanks 1991: 15)

It is important to note that it is highly unlikely that all members of a community participate on an equal basis in this scenario. Rather, Lave and Wenger's research focused on the inherently unequal power relationships related to learning and apprenticeship in communities of practice. Significantly, these communities (of Yucatec midwives, Vai and Gola tailors, US Navy quartermasters, meat-cutters, and non-drinking alcoholics in Alcoholics Anonymous) were made up of individuals with differing levels of expertise and experience. More senior members, who were seen to legitimately 'own' knowledge and learning resources, were therefore perceived to be in distinct positions of power relative to others. They also tended to act as gatekeepers – by implicitly and explicitly choosing who could have access to particular learning resources – and often had the authority to make decisions about how those resources were passed on to others in the community.

Lave and Wenger's discussion of 'legitimate peripheral participation' amongst communities of practice is a particularly useful one for research on environmental education because it emphasises the dynamic and social nature of learning and especially its relationships to identity, power, and change. Certainly, a community like Monteverde cannot be neatly categorised into a single community of practice; it is a community of even wider scope and complexity than the arrangements that Lave and Wenger explore. Nevertheless, both their work and that of many others (cf. Wenger 1998; Bandura 1977; Leeuwis and Pyburn 2002; Keen et al. 2005), usefully highlights the dynamic and social nature of learning. Given the often highly-charged relationships associated with environmental protection and development in many contexts, these concerns about knowledge and power also have clear relevance to environmental education and learning.

As outlined in Chap. 1, researchers have recently begun to explore some of these issues, and especially the ways in which individuals and groups can be involved in learning processes. A recent collection edited by Arjen Wals (2007), *Social Learning towards a Sustainable World: Principles, Perspectives, Praxis*, for instance, provides some important insights into environmental learning not only in formal education, but also in relatively neglected areas such consumer education, corporate social responsibility, community education and cities. The collection is also significant in its attention to the development of the field of social learning across a diverse range of international contexts, including South America, Asia and Africa. This body of work represents a relatively new area of environmental education research, but given that in much international policy '*learning*, in some sense, has supplanted economic growth as the metanarrative and vehicle for bringing about a more sustainable and desirable world for all' (Glasser 2007: 38), it is clearly an important one.

It is for these reasons that I argue that the public sites such as those I observed in Monteverde should be seen as important reflexive spaces where community actors meet, debate and negotiate knowledge – in other words actively engage in learning – related to local environmental management and community development. This knowledge and learning, in turn, served to inform and influence public opinion, to impact upon the implementation of community projects, and to feed back into ideas about the 'appropriate' content and pedagogical orientations of more formal environmental education programmes such as those organised by local schools and conservation organisations. It is in these ways that environmental education in Monteverde – and by extension, potentially in other communities – is deeply embedded within a complex and dynamic 'educational infrastructure' made up of particular institutions and sites for learning, as well as the economic, social and political relationships between diverse community members, and their active negotiations of knowledge.

In Monteverde, the absence of a strong state authority also meant that public spaces were particularly significant as sites for both the distribution of information about local concerns and events, and for negotiation of local development decisions. Rather than relying on a single state authority, local residents have instead had the opportunity to choose to support the work of any of a number of influential local organisations – each with its own perspective on local development, as well as its own range of economic and social interests to promote. The ways in which local residents chose to participate – or not – in these organisations and the fora they organised, therefore, was indicative of wider social and economic relationships within the community, and particularly of struggles over access to resources and the achievement of (sometimes differing) community development goals.

Local Development Issues

In addition to environmental protection and conservation issues, concerns about local development and infrastructure were also increasingly high on local organisational agendas during the time of this research, largely due to the continuing growth of the local tourism industry. Rather than maintaining an almost singular focus on forest conservation, the attention of several local conservation groups was increasingly turning towards concerns associated with population and infrastructure growth. These included problems more typically associated with urban settings, such as water and waste management, as well as the need for improvements to both transportation and communications infrastructure and to the enforcement of existing regulations related to development. In the absence of an effective local government regulatory agency, construction of homes and commercial spaces and the establishment of businesses without legal permits, as well as construction in illegal areas – such as near water sources or in erosion-prone areas – has been commonplace in the region since settlement began. A constant complaint from residents about this lack of regulation and enforcement in the local context was echoed in the commonly-heard the phrase '*en Monteverde no hay planificación*' (literally, 'there is no planning in Monteverde').

Local residents also expressed considerable worry about local methods of water and waste management, especially as the regional population grew and tourism visitation continued to increase. Provision of clean water to the community was not problematic as such, largely because water sources were drawn from high, unpolluted elevations. In 2003, the local branch of the *AyA (Acueductos y Alcantarillados)*, the national agency responsible for water management, was even awarded the government's highest award (a *bandera blanca* or 'white flag') in recognition that local water sources were 100% pure. According to the *AyA*, the award represented its years of work maintaining the system of water collection, delivery and treatment for safe consumption (*Asociación Agua Pura* 2003: 14–15). Local management of waste water post-use, however, was more problematic. There were no co-ordinated regional or village systems for 'black water' (sewage), so the vast majority of homes and businesses used individual septic tanks. 'Grey water' (including domestic run-off from sinks, showers and washing machines, and commercial run-off from local industries such as a cheese factory, pig farm, automotive shops and laundromats), on the other hand, was treated only rarely and commonly ran straight out into gardens, creeks, or the main road. A survey in 2002 showed that approximately 91% of residents believed that a community water treatment system should be established, and were willing to pay for the service (MVI 2002a), but no system had yet been created. Residents, especially in areas of high population density such as central Santa Elena, frequently complained that grey water running through the streets was unsightly and had an unpleasant smell, and tourism promoters worried that the problem would tarnish the community's cherished 'green' reputation.[1]

Throughout Costa Rica, rubbish from households and businesses has historically either been dumped into local waterways, burned or buried. During a lecture on recycling to a group of local students, one local environmental educator succinctly characterised national trends in waste management methods in this way:

> *Ticos have always thrown waste in the water in the belief that they were washing it away, but of course they were really just washing it down to their neighbours... In the 1950s, we buried trash in our backyards or out of the way somewhere, then in the 1960s the state opened landfills. By the 1980s these were getting full, and in the 1990s we finally realised that something else had to be done.*

In the past – when the bulk of solid waste produced by households and businesses was composed of organic materials, and the population relatively small – dumping, burning and burying rubbish was somewhat less problematic than in more recent years. However, these methods of waste management are increasingly dangerous for both environmental and public health now that chemicals and plastics are in common use.

In Monteverde, the Santa Elena Development Association (*Asociación de Desarrollo Integral de Santa Elena*) sought to mediate this problem by establishing a waste collection programme. To participate in the programme, residents and businesses purchased specially labelled plastic rubbish bags from the local supermarket, which were then picked up by the Association's trucks once a week and taken away to a landfill in Puntarenas. The Monteverde Reserve also ran a small-scale recycling programme which collected an estimated 18% of local refuse in 2003. However, many residents were – for reasons of cost or inconvenience – unwilling or unable to participate in either of these programmes, and continued to burn or bury their rubbish as before.

Significant changes to local management of such concerns occurred in 2003 as a result of the establishment of a new municipal council.[2] Prior to that time, the region fell under the political authority of the municipality in the city of Puntarenas, the legislative head of the province of the same name. Monteverde residents felt little

---

[1] Additionally, research suggests that pollution at higher elevations has had significant health implications for communities located downstream (see MVI 2003).

[2] Monteverde is only one of the many rural communities in the country which, due to inadequate transportation and communications infrastructure, had limited interaction with its designated municipal authority. In 2002, the national government passed legislation in response to this problem which created a new category of municipal councils, the effects of which were uncertain at the time of this research. See the '*Ley general de concejos municipales de distrito*' *(número 8173)* – published in *La Gaceta* 10 January 2002.

connection to this provincial head of government, however, partly due to its geographic distance (2–3 h away by bus) and partly because of a perceived lack of interest in local affairs on the part of the provincial government. As a result, few residents paid taxes or expected assistance from municipal officials. Paying taxes to Puntarenas, one resident told me frankly, would be like 'sending money down a black hole'. The first operational municipal council in Monteverde took office in February 2003,[3] and local residents began to re-negotiate and re-formulate local patterns of interaction and organizational responsibility in response to the presence of this new state authority.

For most residents, the ultimate goal for local development was to find a balance between maintaining the ecological integrity of the region, successfully promoting it as a site for tourism, and providing for the needs of local residents, but there were significant differences of opinion about how best to go about doing this in practice. One well-known example of this was a long-running local debate over whether the main road from the capital city should be paved. In order to reach the region, visitors must make a three-hour journey on a narrow, rocky road which winds slowly upwards from the Pan-American Highway to the village of Santa Elena at an altitude of approximately 1,400 m. The Monteverde Reserve lies even further along this narrow dirt road, another 8 km from the main tourist area in Santa Elena, straddling the Continental Divide at between 1,500 and 1,900 m.

Many residents complained that the unpaved road was dangerous for travel both in vehicles and on foot, especially in the rainy season when it becomes a mass of slippery mud. As one author somewhat poetically described it: 'a bone-jarring, muffler-mashing, switchback dirt road' (Honey 1999: 150). Anecdotal evidence also suggested that increasing rates of respiratory problems, especially among local children, were at least partly the result of great clouds of dust thrown up by traffic on the road in the dry season, as well as from increasing levels of air pollution from passing vehicles. Supporters argued that paving the road would benefit farmers attempting to transport fragile produce to external markets, residents in need of emergency medical care, and local transport companies whose vehicles and drivers were continuously rattled by the poor road conditions. Many other local residents, however, expressed concern that a paved road would mean less control over local tourism growth and therefore result in further environmental degradation and social impacts (e.g. increasing alcohol and drug abuse, and the perceived erosion of local social values). Some tourism business owners, and in particular hoteliers, also feared that paving the road would open the community up to exploitation by large tourism companies, and so would endanger local enterprise.

Issues such as this were often at the centre of ongoing public discussions within and among local individuals, organisations, business groups, and the municipal government. These debates were frequently played out in the public spaces – especially ecotourism sites and public meetings – where local residents interacted.

---

[3] A council had actually been elected in the previous election period, but due to a series of problems with funding and organisation was unable to operate and was functionally abandoned.

Community Organisation and the Importance of Public Spaces for Education

Locally-based organisations were especially significant in the social and economic life of Monteverde in 2003 because the community's particular style of organization was characterized – and in many ways defined – by the multitude of NGOs, committees, commissions, and other informal groups that were routinely formed by residents to deal with local concerns or projects. Indeed, interactions in public spaces and events were a central feature of community life. Throughout the year of fieldwork, I attended on average at least two (and often many more) public gatherings each week. These provided useful opportunities for me to contact and communicate with local residents, and to take note of local discussions about community development. In the absence of a strong state authority, these groups played an exceedingly active role in local decision-making and were a vital source of information and debate regarding local environmental management and development decisions. In particular, local organisations relied heavily on public meetings or consultations as a means of communicating information and encouraging the participation of community members in local decision-making. According to one environmental educator in Santa Elena:

> 'There is a culture here of forming committees to deal with anything. Sometimes they are formed and "light the way" for a short while before disappearing, sometimes they stick around for a long time and achieve their goals, sometimes they just don't work at all. There is just something special about the history of this community that makes this possible.'

Meetings and other types of activities sponsored by local organisations thus provided important opportunities for residents to participate in local decision-making, and for local activists to contest knowledge or understandings of local concerns. At one public meeting I attended during the course of my fieldwork, a young Costa Rican resident and tourism industry spokesman joked with the assembled group that 'there are more meetings in Monteverde than in all the rest of the world' – a remark that was greeted by much wry laughter. Despite its joking tone, this passing comment points to the importance of the almost constant stream of public meetings, lectures, workshops and conferences in the life of the community. Previous research also suggests that residents themselves recognised their reliance on local non-state organisations as sources of information about local events and happenings, and that many actively preferred these to more formalised modes of communication. In a survey in 2002, for example, 61% of residents signalled a preference for receiving information about local affairs and events either verbally (individual; face to face) or in meetings (MVI 2002a).

Of course, public gatherings were organised not only to disseminate information and generate discussion, but also to fulfil a variety of more functional purposes. Some, such as annual 'general assembly' meetings, were mainly intended to address the individual administrative needs of each group. Other gatherings included meetings of school boards and parents' committees, church groups, and committees formed to organize festivals or other local events. Commissions and committees of diverse types have also been created by local organizations or individuals with the goal of promoting cooperation between more formally-structured institutions. These inter-institutional projects have been carried out with varying degrees of success in the region since at least the late 1980s, and are an important part of local conservation and development efforts.

Probably the most famous inter-institutional project in the region is 'Monteverde 2020' (MV 2020), an 'organization of organizations' established in the late 1980s that organizers hoped would provide 'a political space for dialog, co-ordination and planning' for ecological, economic and social sustainability within the community (Maroto 1997). The project's name is a reference to its main feature – an extensive series of workshops that invited local residents to examine and discuss perceived changes in the region during the previous 20 years, and to envision what they would like to see happen in the next 20 years. Out of this public consultation several issues of local concern were identified, and proposals for community projects were made. A three-year grant from the Interamerican Foundation was later secured which provided organisers with enough resources to begin a garbage collection program, improve local education provision, and stimulate an awareness of the need for long-term planning in order to protect local environmental, social, and economic resources (Burlingame 2000: 379).

Despite achieving such successes during the first 3 years, however, the project was later suspended. Partly this was because the project's funding came to an end, but also because participants began to realise that – despite all of its good intentions – the project lacked sufficient political power and resources to effectively enforce new policies and decisions (see Maroto 1997). MV 2020 continued to be highly regarded by the majority of local residents during the time of this research, however, because of its emphasis on participation from large sections of the community. Local organisations continued to use similar discourses of participation, collaboration and co-operation to promote community involvement in projects, and to explicitly employ public gatherings as spaces for the dissemination of information and public discussion of local concerns.

Local Styles of Debate and Negotiation

The large number of local organizations and interest groups in the community, and the public fora they held in 2003, provided ample opportunities for observation and analysis of discussions of community development and environmental management in Monteverde. Of particular interest were the competing local discourses of 'sustainable development' and 'development' which were voiced and promoted by diverse organisations and individual actors. Organisations which were categorised by their own members as 'conservation', 'education' or 'ecotourism' groups, and which tended to have high proportions of foreign residents amongst their memberships, more frequently employed a vocabulary of 'sustainability'. Groups such as business or local development associations, in which Costa Rican nationals and powerful local business families were more strongly represented, by contrast, tended to focus more heavily on 'development' as it related to infrastructure development and the provision of services.

Despite these rather fundamental differences in perspective, public discussion in Monteverde was overwhelmingly characterised by statements that reiterated the commitment of organisations and individual residents to promoting co-operative and collaborative projects, as well as democratic and equal participation. At all of the many gatherings I attended, positive statements about the potential for future collaboration between diverse groups and interests were frequently invoked, even when members of the audience were keenly aware of on-going inter- or intra-group conflicts. This diplomatic, and often very elegant, style of speech utterly dominated public modes of communication in the region. Private modes of communication, for example between family members or close business associates, were often a good deal less diplomatic, but by their very nature more difficult to access. Nevertheless, an overall framework of local communication in which diplomatic debate about local issues was complemented by informal communication in private settings was evident. The result of such intensive communication within a relatively small population was that the majority of residents were both keenly aware of local issues and of the projects initiated to address them.

Discourses of Sustainable Development...

One interesting example of this diplomatic style of local communication can be found in a discussion of issues of 'sustainability' during a community meeting I attended in May 2003. The meeting, hosted by the Monteverde Institute (MVI or 'the Institute'), was advertised on posters and in email announcements under the dual-language title '*Monteverde Sostenible*/A Sustainable Monteverde' and took place on a Thursday afternoon in a seminar room at its offices in the village of Monteverde. It was one of more than two dozen workshops and seminars sponsored by the Institute during the year, as part of the organisation's mission to promote education, culture and scientific research, and to offer meetings, symposia, conferences, courses, and informal talks, on cultural, educational or scientific themes for both community residents and visitors.

When it was established in 1986, the Monteverde Institute's founders set out to provide international study programmes in tropical ecology, conservation, agro-ecology, and Spanish. In 1990, however, the organisation broadened its focus further to include sustainable development – and especially the practical implementation of it in the surrounding region. By 2003, many of the organisation's international study courses and public presentations focused heavily on these issues. Study programmes offered during the year included courses on sustainable landscape and architecture, and public health. Students enrolled in the study programmes – mainly from the United States – were allocated a homestay family (usually Spanish-speaking), which provided them with a room and meals in a family home for the duration of the programme (between 2 and 8 weeks). Proceeds from the homestays were used to fund another of the Institute's programmes, known as '*Vida Familiar*' (or 'Family Life'), that provided local families with courses in public health and support in the prevention of domestic violence. Additional projects in 2003 included research on regional water management issues (cf. MVI 2003; Dallas et al. 2001), the annual Monteverde Music Festival, the '*Enlace Verde*' project (the 'Monteverde Greenways' Project which worked to establish conservation easements in the village of Monteverde), and *Finca La Bella* (a cooperative farming project in the nearby village of San Luis).

During our first meeting in January 2003, the Institute's director – a resident originally from the US who first began working for the organisation in 1992 – told me that he viewed the organisation's work as inherently educational and oriented towards meeting the needs of the community. However, the diversity of perspectives represented in the community, he added, often made this a complicated task:

> 'It's really hard to balance the different expectations and the different communities that there are here. It is the Institute's philosophy to be a meeting place of different ideas, different people, different communities, different cultures. So we're a place where you hear the voices of biologists, of local *campesino* farmers, of ecotourist entrepreneurs… and sometimes that's a cacophony and sometimes [laughing] hopefully you can build consensus around it. It's not easy to be as open as we are to hearing different points of view because we also then get criticised by everybody. But, you know, that's part of what we're trying to do, is to be a place where there can be a forum for discussion where we are dealing with the very difficult issues that we have in Monteverde, and addressing population growth, environmental degradation, the ups and downs of the tourism industry…'

Despite the common use of such inclusive language by both the director and other staff at the Institute, however, audiences at the organisation's meetings were most often composed of foreign residents and visiting researchers, for many of whom English was their first language. The majority of the audience for the gathering in May 2003, for instance, was made up of students from the US who were taking part in one of the Institute's study programmes. The remainder of the assembled group was made up of a dozen or so residents – Costa Ricans, foreign settlers and members of the Quaker community – who were already involved in the Institute's work, and myself. As was the case with the majority of meetings that I attended at the Institute and elsewhere during the year, both the panel of speakers and the audience contained a mixture of native English and Spanish speakers, so the staff provided simultaneous translation.

For the purposes of this particular meeting, the director had proposed three specific questions as the focus of the discussion: 'How sustainable are we now? Where do we want to be in the future? How do we get there?'. He had also invited a group of panellists to discuss 'sustainability' in relation to their particular areas of expertise: the region's newly-elected mayor would discuss governance and regulatory issues, a long-time resident biologist from the US would frame her comments around conservation efforts, a representative from a local coffee cooperative would address local agricultural issues, a representative from a local tourism interest group would talk about the local tourism industry, and one of the Institute's own

Local Styles of Debate and Negotiation 119

researchers would provide results from the Institute's work on local public health and human development.

As the discussion progressed, it became clear that while the panellists largely agreed on the underlying principle of sustainable development – namely the need to balance environmental protection with economic and social development – there were also significant differences in their perspectives regarding what sustainable development in the community might mean in practice. While the diplomatic style of communication, which continuously evoked the need for cooperation and collaboration, in some ways obscured these fundamental differences in approach, key questions were raised about how the term 'sustainable development' was meaningful locally, and how diverse community members variously used it.

The first to speak was the mayor – a Costa Rican resident born in the region – who talked in some depth about the work of the municipal council during its first few months in office. He divided his assessment of the community's current state of 'sustainability' in terms of issues surrounding conservation, health, local industry, transportation infrastructure, and youth programmes. In his view, central to work in conservation was the encouragement of a 'culture of care for natural resources' (flora and fauna), especially through effective water and solid waste management. He stated that health services were improving, but that continued vigilance would be required to protect residents from public health issues (such as drug addiction and the spread of HIV/AIDS) resulting from increased tourism. Speaking partly from his position as a partner in a local ecotourism enterprise himself, the mayor argued that local industry was already highly sustainable, but that businesses would also need to improve waste and water management. Also important to the future of the community, he concluded, would be programmes for young people which would encourage them to be healthy and active members of the community. Overall, the focus of the mayor's comments was on increasing and improving planning – of natural resources, of health, of water and waste, and of educational opportunities and community programmes for young people – as well as the regulation of development. In particular, he talked about the need for community groups to work together to create a regional regulatory plan (*plan regulador*) that would ensure 'sound planning for sustainable growth and development'.

The second presenter – a long-time resident from the US, biologist, and founding member of the newly-established Costa Rican Conservation Foundation – provided a rather different perspective on local sustainability as it related to conservation efforts. In fact, she did not use the term at all in her presentation, and implied instead that local conservation was not yet 'sustainable' and would only become so when larger areas had been given legal protection. To begin, she gave an overview of new research about local conservation successes and losses, and accompanied this with a slide presentation showing familiar images of endemic flora and fauna in protected forest areas as well as scenes of regional deforestation caused by agriculture. She also described increasing concerns about forest fragmentation on the Pacific Slope and its consequences for local species diversity. These changes are important to Monteverde, she added, because species like the Three-Wattled Bellbird and Quetzal – that can only be found on the Pacific Slope, and outside of already-established protected areas – are being seriously effected. Decreases to, or loss of, these populations would have significant impacts locally because they are important both to scientific study and the tourism industry. Overall, the talk presented an image of fragile local ecosystems in serious danger from human activities, and the speaker implied that the only way to 'sustain' these was by extending legal protection to even larger areas.

The third panellist to speak was a representative from a local coffee cooperative, *Cooperativa Santa Elena*. Humberto Villa is a dynamic man with a long history of involvement in local conservation, education and development issues, during which time he has worked for several local NGOs (see Chap. 4). Although he had been invited on this occasion to make remarks specifically about the local agricultural sector, he began by speaking generally on the topic of community sustainability, describing how the local government, NGOs and local businesses had already formulated their own ideas about development and sustainability, and were working to protect large forested areas. He then outlined another vision of 'sustainability' in the community. Rather than talking about regulation or protected areas, he spoke instead of the need to promote self-reliance and the sustainable development of local resources, so that the community is not dependent on international or national trends. Specifically, he mentioned what he called a 'worrying dependence' on tourism as the only source of local income, on the ability of large landowners to protect local forested areas (specifically the Monteverde Conservation League and the Tropical Science Center), and on the use of imported products such as medicines and food which could be produced locally. He argued that depending solely on tourism for income was particularly worrying because approximately 80% of visitors to Monteverde are from the United States, and so what happens there could have a huge impact locally: *It would be the same kind of mistake as a coffee farmer who agrees to sell all of his harvest to one client. What happens to the farmer if the client suddenly decides not to buy?* He suggested that greater efforts should be made instead in the development of local production and management:

> *This region has a long history of richness in shared experiences and cultural exchange, beginning with the arrival of the Quakers and extending to today's student groups, volunteers and tourists. At the same time, it is dangerous to depend too heavily on outside sources for income, conservation, or other benefits. To avoid being vulnerable and to make local development sustainable, we should focus our conservation and development efforts within the community itself.*

In yet another perspective on 'sustainability', the administrative co-ordinator for the local Chamber of Tourism spoke about the need to manage local tourism in harmony with nature:

> *What is most important is that local businesses see that sustainability requires not only economic sustainability, but also environmental sustainability… If what people come to Monteverde to see in the first place is destroyed, the community will lose all the economic benefits.*

The key to achieving sustainability, in his view, was to manage and effectively regulate the tourism industry in such as way that the industry continues to prosper while any negative impacts are minimized. The history of the development of tourism in Monteverde, he commented, has been an 'abnormal' one in some ways because tourists – first scientists and later others – began to arrive long before any infrastructure had been put in place or plans made to take care of them. In order to manage the industry effectively in the future, he concluded, the government, businesses and other local organizations would need to continue 'meeting and talking as a community' and working together with 'a single vision of Monteverde's future in mind'.

The last presentation of the day came from the Institute's academic programmes co-ordinator, a young woman from the US who had arrived in the community in the previous year to take up the post at the Institute. Her talk was a presentation of the data gathered during a large-scale survey conducted by the Institute in the previous year (see MVI 2002a), as well as information less formally gathered in partnership with the local health clinic. She began by describing international policy understandings related to the 'pillars of sustainable development' (economic development, social development and environmental protection) and the 'fundamental conditions of sustainable development' (outlined during the Rio+10 Summit in Johannesburg, South Africa in 2002), and which she then used as the framework for her assessment of the region. According to the information gathered, she told the audience, Monteverde rated relatively well in terms of access to clean water, housing, health care and energy, and the protection of local biodiversity. Problem areas, however, were identified in terms of sanitation (98% of grey water untreated; garbage collection limited), poverty eradication (increasing income disparity with the growth of the local tourism industry), technology transfer (limited public access to internet and computing resources), human resource development (up to 50% of local students do not complete secondary school; high rates (18%) of teenage pregnancy), education and training (limited adult or further education; limited opportunities for distance education), and women's empowerment (lack of services for victims of abuse; as of yet no data regarding the impacts of the increasing entrance of women into the local workforce; limited support for working women/mothers). Of all the presentations on the panel, this last one used perhaps the broadest frame of reference for 'sustainability', by encompassing a variety of environmental and human factors. Interestingly, it was also the only talk to make direct reference to international policy definitions, although all of the previous discussions of 'sustainable development' made at least indirect reference to similar, if not identical, policy concerns.

While it would be far too simplistic to suggest that the perspectives presented by these individuals could be easily attributed to particular groups within the community, there were nevertheless strong parallels between the diverse ideas expressed during the meeting and on-going discussions that I took part in within the community as a whole. The biologist's call for the establishment of larger protected areas, for example, was also vocalised by many natural scientists and conservationists who resided in the community and whose work was oriented towards strict habitat protection. This understanding of conservation, however, stood in marked contrast to comments by community actors – such as the representatives from the coffee cooperative and the Chamber of Tourism – who argued for multiple-use schemes and the greater involvement of local producers and the business community in conservation efforts.

... and Competing Discourses of Development

Of course, not all community members approved of, or even commonly used, the phrase 'sustainable development', and there were those in the community who instead argued for quite different kinds of answers to questions about local development and environmental management. There were many community actors who spoke instead of the need for 'development' – by which they referred to a perceived need for greater service provision and improvements to local infrastructure. One of the most visible groups which actively promoted this perspective was the Santa Elena Development Association, an organisation which was founded by Costa Rican residents of the village of Santa Elena in 1975. Similar development associations can be found throughout the nation, particularly in rural areas where the state has historically provided only limited services. As such, the group's work focused on organizing its members to deal with practical concerns such as construction and improvements to roads and bridges, household and business waste collection, and postal services. The organisation's substantial assets included heavy road maintenance equipment, transport vehicles (especially for the waste collection programme), and three buildings in Santa Elena (one served as their own office space, and the others were rented out to the Red Cross, the police, and the municipality).

Discussions that I observed at the organization's general assembly meeting in July 2003 illustrated an alternative understanding of 'community development' to that described at the Monteverde Institute above. The meeting was formally advertised by broadcasts from a local pick-up truck with a loudspeaker in the back, and also informally by word-of-mouth. It was held at 11 o'clock on a Sunday morning in the parish hall (*salón parroquial*) in Santa Elena. As members and visitors arrived, they were asked to sign in a large ledger and then each received a name tag which also listed their affiliation to the organisation.[4] By the time the meeting began, the room was filled with several hundred people – men, women and children of various ages. The majority of these were Costa Rican residents and native Spanish speakers, although several members of the original group of US Quaker settlers were also in attendance. The board members themselves were formally seated at a long table on a raised stage at the front of the room. Each one was briefly introduced to the audience by the current president of the executive board, an older male member of a prominent local business family, and they made reports on the year's progress in their respective areas of responsibility.

Topics of discussion as the meeting progressed included reports on current projects, plans for the construction of a multi-purpose community centre for local youth, and progress reports about current road improvement projects and maintenance of the organisation's equipment (including tractors, trucks, and storage sheds). Reports, agenda items and motions introduced by assembly members were followed by time for anyone in attendance – including members, board members and visitors – to

---

[4] As an observer, my designation was '*Visitante – Voz Sin Voto*' (Visitor – Voice without Vote).

respond or ask questions; a time-consuming process which lasted for several hours.[5] While some issues for discussion were mandated by assembly by-laws, including new board elections and financial reports, others arose out of specific audience concerns. In one instance, a member asked for assistance on behalf of a fellow member who had a concern regarding one of the Association's vehicles. Apparently, several years before a truck had been bought in the member's name and he had donated its temporary use to the Association. However, he now needed legal papers to show that the truck was, in fact, his property. This concern reflected a time in the not-so-distant past when actions by the organisation were taken in a relatively informal or personalised manner – usually through a verbal agreement. After a short lunch break – during which *arroz con pollo* (rice with chicken, a typical Costa Rican dish), salad, crisps and fruit juice was served to the entire assembly by a large crew of women working in the kitchen attached to the meeting space – the meeting continued to proceed in great detail through each of the motions and agenda items. The discussion then moved on to elections for the new board of directors at 5 o'clock, and the formal meeting itself finally finished around 8 o'clock that evening.

As in the case of the Monteverde Institute meeting, discussion at the Association's general assembly was characterised by the typically diplomatic local style of dialogue, which emphasized cooperation among community members and open debate about local decision-making. Strikingly, this was despite the fact that there were observable tensions among individuals within the membership, as well as evidence of both past and present disagreements. It was a rather formal and solemn event overall, a feeling that had been underscored by the opening comments of the moderator, Humberto Villa (previously a panellist at the Institute meeting). He presented a short talk about the 'rules of the road' for the gathering, in which he asked that all participants speak and behave towards one another with respect, make constructive criticisms, and try to keep comments brief – in order that everyone might participate and the meeting could be as short as possible. He concluded by asking that the assembly 'keep in mind the future and what we want for the community, when you make comments or come to decisions today'.

These comments, I was later told by Association members, were largely in response to on-going conflicts among the membership. As an indication of these concerns, the president of the board, for example, had also informed the audience at the beginning of the meeting that the proceedings were to be recorded in order to avoid undisclosed 'problems like we have had at meetings in the past'. Prior to the meeting, I had also talked about both the Association's work and the upcoming gathering with members of a women's advocacy group in Santa Elena. One of the women in the group, who was also an Association member from a Costa Rican family with long roots in the community, recounted the way in which a sub-group within the organisation was quietly attempting to convince fellow members to vote the current

---

[5] Friends in the community had told me that assembly meetings like this one often ran long, and the level of detailed discussions and attention to local social conventions (which essentially require everyone to have a say) did indeed make the process quite extended.

president out of his position, and to replace the entire board with members affiliated with their own local political party. Other women in the group appeared unsurprised by this news, and commented that there was always 'some sort of power struggle' happening in the organisation. They all agreed, however, that if any under-handed tactics were being used, they were inexcusable.

Despite such predictions of conflict at the meeting, however, during the gathering itself the Association's members employed a vocabulary of collaboration and mutual respect that was largely identical to that used in other public gatherings in the community. The noticeable difference in this gathering was not in regards to the vocabulary, therefore, but in terms of the topics and issues being discussed. Concerns about road works projects and decisions about the use of organisational resources such as buildings and vehicles, for example, demonstrated that the Association's community development agenda was rather different to that of other local organisations. Namely, underlying the Association's projects was a much stronger focus on providing a better quality of life to residents through the creation of, or improvement to, local infrastructure and services – a strong contrast to the 'sustainable development' efforts promoted by groups such as the Monteverde Institute.

Participation in Local Organisations

By setting up a contrast between the work of these two diverse groups within the community, and making distinctions between their approaches to community development, I do not mean to suggest that they operated in complete separation from one another, or in an environment that was hostile to collaboration. On the contrary, as the discussion above also shows, there were many individuals participating in, working within, or enrolled as members of, several different local interest groups. Humberto Villa, for example, has appeared in this book in his capacity as an environmental educator (Chap. 4), as a representative of the *Cooperativa Santa Elena* at the Monteverde Institute meeting, and latterly as a member of the Santa Elena Development Association. Indeed, opportunities for collaboration such as these were strongly emphasised in public discussion, while disputes between individuals and organisations tended to remain hidden. In many cases, conflicts based on differences in perspective or personal disagreements were usually only visible in the public sphere through a *lack* of co-ordination between groups or individual participation in projects. In this context, the decision to participate in specific organisations or to collaborate in projects can be seen as a strategic choice by both permanent residents and visitors, and thus also as having a strong impact on processes of 'community development' in its widest sense.

Through time spent in a wide variety of meetings and other kinds of gatherings throughout the year, and via candid conversations with the actors involved, it was clear that there were often explicit reasons – ideological, social and practical – why residents chose to participate (or not) in particular local organisations. Firstly, and perhaps most obviously, both short- and long-term residents were most likely to become involved in groups whose aims and goals – in terms of infrastructure development or 'sustainability', for example – they believed to be the most appropriate ones for their community. Secondly, while local NGOs generally advertised their events or projects as open to participation by any interested individuals, there were significant social divisions within the community that influenced the likelihood that people would indeed join in. These revolved around the use of language (Spanish or English) and vocabulary (especially in terms of differing educational levels) among particular groups, as well as local settlement patterns. Residents of Santa Elena in 2003, for example, were mainly Spanish-speaking Costa Ricans, while residents of the village of Monteverde had largely come from other countries (most commonly the US). Even when these kinds of issues did not dissuade residents from taking part in the public discussions initiated by a particular group, more practical issues such as the high costs of and limited access to local transportation, access to the communications media employed to publicise events (e.g. emails, posters, loudspeaker announcements), child care needs, long working hours, or limitations of citizenship often had an impact on the ability of individuals to participate in some groups.

In terms of individual membership in local organizations and participation in events, the orientations and activities of specific groups often attracted members from particular sectors of the community. Groups like the Monteverde Institute which focus on 'sustainable development', for example, tended to derive their memberships from among the more recent arrivals to the community – from other parts of Costa Rica or the rest of the world. Many of these individuals came to settle in the community precisely because of its reputation for successful conservation, and so were likely to be pre-disposed to become involved with local groups with a particular perspective of environmental management and development. Many members of these groups in 2003 were natural or social science researchers, with the result that these organisations tended to emphasise research-oriented projects and modes of problem-solving (cf. MVI 2002a, b; 2003), and discussion that was heavily informed by contemporary international policy debates. Participants in local 'development'-oriented groups, on the other hand, tended to be more heavily involved in the profitable local tourism industry (particularly members of influential local Costa Rican families), and therefore had a significant economic interest in the continuing improvement of infrastructure and the growth of local business. Additionally, organizations that had a majority of US or foreign members were more likely to advertise and conduct events in English. As shown in the case of the 'Sustainable Monteverde' meeting, many organisations did make an effort to provide simultaneous translation, but local residents whose primary language was Spanish were often uncomfortable asking for assistance.[6] Local groups with predominantly

---

[6] I came across this issue myself when arranging to give a presentation about my research to the community at the end of my fieldwork stay. Following the presentation a number of people – some of whom had not even attended the session – thanked me for having conducted it in Spanish. This reflected a much wider problem of communication of research results to the community and resentment on the part of Spanish-speaking residents who felt that access to information and research results was too limited.

Spanish speaking leadership and memberships, including the Santa Elena Development Association and the municipal council, on the other hand, were usually conducted in Spanish and without translation. As a result, attendees are more likely to be native speakers and fluent foreign residents.

Communication of information across the community was also impacted by linguistic and infrastructure barriers. Organisations that relied heavily on email lists to advertise events, for example, often failed to reach community members who did not have regular access to the internet. According to estimates at the time, less than 19% of households in the region had a computer in the home, and even fewer had access to an internet connection (MVI 2002a). At the same time, access to information was not always predicated on access to technology. In the Monteverde region, as in other parts of Latin America, local events (such as the Santa Elena Development Association's meeting above) are often announced via loudspeaker from the back of a passing pickup truck. This egalitarian advertising method was intended to ensure that messages were heard by all residents, regardless of educational level or access to other media. A few years previously, however, complaints from the (mainly foreign) residents of the village of Monteverde about the noise produced by this practice led the announcer to cut the village out of his route, potentially causing residents there to miss out on some local news and announcements.

Geography and the local economy played a significant part in individual participation as well. Firstly, there were only limited spaces in the community which were suitable for large meetings – one each in central Santa Elena and Cerro Plano, and two in the village of Monteverde. These belonged to the Catholic Church, the Monteverde Conservation League, the Monteverde Institute, and the Monteverde Reserve, respectively. The meeting spaces were available for use with permission from the owners, either for free or with a small user fee, depending on the circumstances. These spaces were often in use by the organisations themselves, however, and access to them depended on the willingness of owners to share the space. Community members who wished to use these spaces also had to consider their relative accessibility to fellow residents. While young and able-bodied community members could easily walk from village to village for work or leisure, the costs of other methods of transportation could be prohibitive for older residents or those with health concerns. The most common forms of transportation included cars, motorbikes and quad bikes, but estimates at the time suggested that 45% of area residents did not own their own means of transport (MVI 2002a). In terms of public transport, one twice-daily bus service ran from central Santa Elena to the parking lot of the Monteverde Reserve, at a cost of between 500 *colones* (residents) and 1,000 *colones* (tourists). Taxi services (both legally registered and 'pirate' taxis) were also readily available, but a single trip between Santa Elena and the Monteverde Reserve cost anywhere from 1,500 to 2,000 *colones* (£3.50–£4.50).

Such high transportation costs discouraged residents from travelling long distances for meetings, and they were therefore more likely to attend events near their own homes. This was especially the case for gatherings conducted during the evening hours or in the rainy season, when travel along the muddy, unpaved main road was particularly difficult. Parents with young children were also less able to attend evening sessions because of child care issues, while many others were not available for daytime sessions because of work commitments. (The nationally-mandated standard for full-time employment in Costa Rica is 48 h per week, and workers in the tourism industry frequently work even longer, and often irregular, shifts in order to meet client demand.)

In addition to simply attending meetings and events, opportunities to take on leadership roles within local organizations were also sometimes limited for local residents. In the case of the new municipal council, residents were required to be Costa Rican citizens in order to either vote for, or work within, the council. In practice, this prevented many foreign residents from taking formal roles in local or national government, although many were active and influential in informal ways, including campaigning for conservation issues through national media or contacts within government ministries. Some community members claimed that the inability of non-citizens to participate in formal state structures was one of the major reasons behind the existence of the region's numerous NGOs. While leadership in these local organizations had no restrictions in terms of citizenship, selection for those roles was nevertheless often heavily impacted by nationality, educational level, socio-economic status, and family or business relationships.

It is in the context of these complicated ideological, social and practical limits that local residents negotiated their participation in influential local organisations and, in turn, gained access to and took part in the continuing formation and re-formation of meanings and practices of local environmental management and community development. In doing so, they also took part in dynamic educational processes related to environmental and development issues in the community. The resulting knowledge and awareness – along with understandings of complicated local relationships – was strategically employed by many individuals and organisations to gain access to local resources (both natural and human) and to achieve particular community development goals.

The spectrum of local perspectives on types of development – ranging from 'sustainability' to 'infrastructure development' – promoted by local organisations were therefore part of continuous processes of discussion and decision-making among community members about both how best to 'develop' the community, and about who (in terms of organisations or individuals) was best suited to manage the development process. While there was general agreement in the community that a balance should be found between the further development of infrastructure (especially for tourism) and the protection of local resources, negotiations of this balance were underscored by long-standing tensions in the community regarding who had access to, and control over, the protection and study of local resources.

Community Development: Ecotourism vs Environmental Protection?

Public debates about styles of community development were in this way interwoven with more practical concerns such as access to local resources. These included not only the region's material and economic wealth, but also its biological diversity and international reputation for conservation and research. Access to such resources locally most often relied on connections to local conservation initiatives and/or the ecotourism industry. Monteverde is internationally marketed – by both groups – as a unique site for learning about 'green' topics such as rare endemic flora and fauna, and for opportunities to have a certain kind of learning experience (i.e. of being in the cloud forest or of seeing a rare bird).

While in theory, these groups supported one another's efforts to protect local ecosystems and livelihoods, conflicts did sometimes arise surrounding competing claims to use and disseminate knowledge about the local environment. These claims were often central to the establishment of business enterprises, individual careers and organisational reputations. In the case of local conservationists and researchers, knowledge of the local environment was most often used to further the pursuit of scientific understandings of the natural world, and to promote the strict protection of ecologically sensitive areas. However, local scientific researchers also used their knowledge of local ecosystems to attract students to study tropical ecology, biology or botany. For a range of personal and employment-related reasons, relatively few of the researchers working in the region resided in the community year-round, and many split their time between Monteverde and a home institution (usually a foreign university or research institute). Many of the dozen or so researchers living permanently in the region reported that they were able to make a living only by relying on more than one source of income – such as organising international study courses, running small enterprises out of their homes (such as selling photos of popular local wildlife to tourists), writing guide books or other publications about their specialist topics, or working on short-term consultancy contracts. As one resident US biologist who settled permanently in Monteverde in the 1980s explained:

> 'There are a lot of biologists who have been to Monteverde and return periodically … and a lot of them will be here in June, July and August, or at least part of that time. Some of them have been coming here for fifteen, twenty, twenty-five years, but they depend on the income from their teaching jobs in the [United] States. They just can't cut it all off and move here… The ones you see that live here are the ones that have actually done that, though… you know, "we're finished with academia, let's go to Monteverde and we'll figure out a way to make a living once we get there". That's what we did. And it's been tough at times.'

There were also relatively limited stable employment opportunities in either state or private conservation organisations for Costa Rican researchers and conservationists, so very few were able to live and work in the region. During a conversation about the role of visiting researchers in local conservation efforts, the same resident US biologist commented:

> 'I think it has certainly been useful through the years to have a core of professional biologists here. Very few communities in Costa Rica have that… I mean, they have biologists who work throughout the conservation areas, but they're not very well paid and that's probably not the job you want if you're a professional biologist. You may take that job because you need a job, but that's probably not going to be your goal from the outset because it's probably not going to get you much real professional advancement and not a very good salary… In general being a Costa Rican biologist is not well-rewarded, in any way. That's not to say that Costa Rican biologists are not involved in environmental efforts in Costa Rica,

because they are. For the most part they're a pretty young group and they're pretty active, but they're mostly located in the Central Valley because that's where the jobs are.'

Despite the financial difficulties and geographical isolation inherent in doing research in Monteverde, however, it continues to attract hundreds of students and researchers each year. For many, work conducted in Monteverde is the foundation for building a reputation within international research and conservation arenas.

Similarly, ecotourism business owners use their knowledge of the local environment for specific purposes, such as attracting clients, supporting local conservation, and providing learning opportunities for both residents and visitors. Many owners I spoke to in 2003 expressed a keen interest in promoting local environmental protection and maintaining the positive reputation of the community, both for their intrinsic value as well as a means to sustain the local tourism economy. The most directly 'environmental' or conservation-oriented of local attractions were a frog exhibition, a serpentarium, a butterfly garden, an orchid garden, and an 'ecological' farm.[7] Each one offered educational experiences related to local flora or fauna, and promised knowledgeable guides and tours in multiple languages. In addition to these, there were also many other tourism businesses which were less directly oriented towards environmental learning, but which likewise utilised visitors' interest in the natural beauty of the local environment as a marketing tool. These included numerous small, often family-run enterprises that offered tours of local destinations on foot, by mini-bus, or on horseback. Many of the larger hotels, and especially those located close to the Monteverde Reserve, also maintained walking trails through private forested areas for guests and visitors.

The most popular, and profitable, of local attractions offered 'bird's eye' views of the forest by riding along on zip-lines suspended in the forest canopy. The first company to begin offering these kinds of tours in Costa Rica – The Canopy Tour – opened in Monteverde in the 1990s, and by 2003 there were three other companies offering competing services. Competition between these companies can be extremely stiff. The Canopy Tour (later re-named 'The *Original* Canopy Tour'), for instance, was granted intellectual property rights for its suspension system in 1998, and in 2003 was in the process of suing local competitors for copyright infringement. As there were already at least 50 other canopy tours operating nationwide by this time, a successful suit would have serious ramifications for the tourism industry as a whole.

According to one local ecotourism business owner – a Costa Rican who was born and raised in the region – the local industry experienced a particularly big boom between 1981 and 1988, and had continued to grow steadily since then. Indeed, during 2003, I observed several new businesses – including the newest forest canopy

---

[7] In Spanish, the *Ranario* (Frog Pond), *Serpentario* (serpentarium), *Jardin de las Mariposas* (butterfly garden), *Jardin de las Orquideas* (orchid garden), and *Finca Ecológica* (ecological farm).

tour and a new insect exhibit – under construction in Santa Elena. Such ecotourism or educational tourism destinations were important to the ways in which the community marketed itself as a tourism destination, as evidenced by the vocabulary of 'environmentalism' that was characteristic of promotional tourism materials. Even local hotels and restaurants often labelled themselves as 'eco-friendly' or used vivid paintings of local wildlife on their exteriors to attract visitors. The Costa Rican tourism industry in general, and business interests in Monteverde particularly, have been especially successful at the marketing and promotion of 'green' or eco-tourism on the internet. In 2003, there were already at least three groups in Monteverde maintaining websites with photographs of local attractions, descriptions of offerings by local hotels, restaurants and tour companies, and links for online bookings, and there are many more today.

The extent to which local businesses actually engaged with environmental management and implemented sustainable business practices varied widely, however, and this was almost entirely self-regulated. All local businesses must adhere to national environmental legislation, but enforcement has been limited both because of the geographical isolation of the community and its relative autonomy from the state. As a result, local business owners in Monteverde have tended to make individual decisions about environmental practices such as water and waste management, energy use, and choice of building materials, among other things. There have been a few collaborative efforts in the community to support environmentally-friendly business practices, however, as well as to co-operatively promote the industry. In 2003, two tourism business groups were engaged in building collective projects with local owners. Organisers of both groups, however, claimed to have difficulty in gathering broad support from the local business community.

One of these groups – the Chamber of Tourism (*Cámara de Turismo*) – was formed in 2000 and operated under the motto '*Monteverde Para Siempre*' (Monteverde For Always). This motto was linked to the group's overarching goal of helping the community to develop 'sustainably' and to maintain its attractiveness as a tourism destination, its plentiful sources of employment, and its peaceful and secure way of life (cited in *Asociación Agua Pura* 2003: 22). During 2003, the group's main project was the creation of a new regional visitor's centre. The idea originated from within the group's membership, but it was being planned in collaboration with the new municipality, the *AyA*, and the Monteverde Institute. Organisers hoped that the centre would further strengthen the community's image as an ecologically-aware community, and also that it would act as an example for other communities to follow.

According to the group's administrative co-ordinator, a young Costa Rican belonging to a prominent local family, the Chamber of Tourism was also working to encourage local businesses to take an active role in the 'sustainable development' of the community. The group's members, he told me, recognised the importance of maintaining the economic feasibility of local tourism: 'The local economy is based on tourism, so if we lose that the community loses everything'. However, the Chamber had only 33 member businesses on its rolls in 2003, representing only a relatively small proportion of the 52 hotels and 32 other tourism-related businesses in the community. Getting people to co-operate had proven difficult, the co-ordinator commented, because of the wide range of opinions and attitudes among local business owners:

> Some of them see membership as a tool for getting specific projects done for the benefit of their own businesses, and others see it as a way for the business community to work together and make a positive contribution to community development. Many members have the attitude that 'If the community does well, then so will I', but others tend to think that 'If I'm doing well, it doesn't matter how everyone else is'.

Competition and Cooperation in the Local Ecotourism Industry

Gaining broad support for such co-operative efforts among the local business community was also problematic because of the diverse kinds of enterprises operating in the region and their intense competition for profits. According to a survey of 93 local tourism businesses by the Monteverde Institute (MVI 2002b), 76% of local tourism business owners identified themselves as Costa Rican nationals, with the remaining 24% originally from the US, Europe and other parts of Latin America. Eighty-three of the tourism businesses surveyed began their operations between 1986 and 2002 (simultaneous with the region's highest period of population growth) and so were in some senses still 'newcomers' to the community. Individual owners responding to the survey also categorised their enterprises in differing ways, for example as pertaining to either tourism (44.6%), ecotourism (26.1%), education (3.3%), educational tourism (9.8%), for-profit business (4.3%), agriculture (2.2%), or the service sector (9.8%). This diversity among local tourism businesses also resists simple categorisation or cooperation according to the type of services offered. Among local hotels, for instance, as few as 10 or as many as 20 rooms may be available to accommodate visitors, with prices ranging from an inexpensive room in a *pensión* in Santa Elena for £3.50 per night to a more up-scale hotel room near the Monteverde Reserve for as much as £65 per night. As the community has continued to grow and gain popularity as an international tourism destination, competition for tourist earnings has intensified. This greater competition has had particularly strong impacts on businesses which re-invest a portion of their profits in community programmes. As one local ecotourism owner commented in 2003: *A tourism business should be able to provide enough money to pay employees and operating costs and also to invest in conservation or educational projects. There are some in the community who only put the extra profits in their own pockets, but that's not the way it should be.*

Another local businessman I spoke with, the founder and owner of the Butterfly Garden, located in Cerro Plano, also expressed serious concern about the growth of the tourism industry and increasing competition. David Adams originally came to the region from the US in 1978 as a self-described 'burned-out academic biologist', at which time he bought a 25 ha farm for study and small-scale agricultural production. In 1990, he decided to take advantage of the rising number of visitors arriving in the region, and to use his background in biology and education to open his own business.

It began with only an information centre and a single garden, but by 2003 had expanded to include three more butterfly 'gardens' enclosed in wire mesh (each containing plants and butterfly species representative of those found at a particular altitude), and several other insect exhibits, such as a memorably large 'ant farm' for a leaf-cutter colony.[8] Guided informational tours – in either Spanish, English, French or German – were included in the entrance fee. Although tours of the exhibits were not compulsory, David encouraged visitors to take them because:

> 'Most people are raised in cities these days, they don't see much in nature... they're not trained to, unless it's the size of an elephant... So by themselves they're not going to see much. And then, we really like that the message they take away about them [about the butterflies] is not that they were *pretty*, but that they learned something'.

In addition to providing educational experiences for tourists, the garden also offered free entry to local students who visited during the low tourist season. David told me that relatively few of the local schools took advantage of this offer, although the private schools tended to visit more frequently. When they do come, he added, it is always because an individual teacher has taken the initiative to arrange it and not because there are established programmes with the schools. When I asked David to tell me more about his personal definition of 'environmental education', he replied:

> 'I think you should just not define it.... I mean there are certainly things that are *not* environmental education... I mean, it teaches people about nature and man's relationship to nature, so I would prefer it myself to be more about topics like biodiversity, but that's just my point of view. I'm sure that most people would be more human-centric, you know, anthropomorphic... 'oh, well, environmental education is protecting our water'. But when you get at it, it's water so that *we* can drink... environmental protection is to protect the forests so that *we* can get wood from it or whatever. I would like to see more environmental education done so people see the value – the existence value – of other species. Then it's for it's own sake... which is definitely philosophical and spiritual and all that... but that's a hard goal in developing countries... it's a hard goal in Latin America.'

David was only one of many people in the community who expressed concern about the nature of environmental education in Monteverde. Some residents worried specifically that too many people in the community were more motivated by a desire to make money in tourism than to protect local resources. David believed that issues like waste disposal and water treatment should be a matter of concern for the whole community:

> 'Before it didn't matter because so few people lived here... hundreds not thousands. The community, sooner or later, is going to have to get together. Unfortunately, I don't think the eco-tourism sector are going to be the leaders. Most of the people making the big bucks here aren't 'eco'.... The hotel owners that by far make the highest profit margin are just business people.'

---

[8] This was a five or six foot tall rectangle of plexi-glass that showed the colony in cut-away, and was reminiscent of the small scale versions manufactured for children to observe at home.

Fellow local ecotourism business owners expressed similar doubts about the future of local conservation and educational efforts. Rafael Carazo, owner of the local Orchid Garden, was also highly critical of local businesses: *Some of the hoteliers should know better, because they were the ones who were taught in the very first environmental education programmes so many years ago.* Rafael was himself also part of these efforts a decade before when he spent 2 years co-ordinating educational programmes at the Monteverde Reserve. At that time, he recalled, the reserve was working in a large geographical area – including the settlements of Cerro Plano, San Luis, Monteverde, and La Lindora – and they ran programmes about a variety of issues, but he felt that there was relatively little support for the projects and only limited resources to draw upon. In the end, he decided that his real passion was for research, and he left the reserve to open his own ecotourism business.

When we spoke one afternoon in 2003, he was still running that enterprise, an exhibition of endemic orchid species located along the main road just outside of central Santa Elena. The site itself, which sits at the bottom of the downward sloping roadside, is very small, but was filled with a huge variety of orchid species bound carefully to trees or other, sturdy plant growth.[9] For a small fee, visitors received a guided tour of the grounds that provided detailed information about particular species, their adaptations, and links to local wildlife. Although only 15–20 people visited the garden each day, Rafael said, he was happy doing this kind of work because there was plenty of time to pursue both his research as well as to participate in educational and ecotourism activities. When there were no visitors, he worked with volunteers and staff to care for the plants, ensuring that they were properly labelled and secured. More than anything else, he told me, his business was a way of sharing his passion with others:

> 'What is so amazing about orchids, and what I try to convey to visitors, is their links with other kinds of life. A commonly-heard example of this is the adaptations of hummingbird's beaks to suit the shapes of certain orchid species. The diversity of Costa Rica's orchids is unequalled in the world – there are 1,500 species in Costa Rica, of which 500 can be found in Monteverde... and there are more being found every day that haven't even been named yet.'

During the many years of his work in Monteverde, Rafael also published widely in academic journals, and he was the author and illustrator of the most popular field guide to the orchids of Costa Rica, which could be found at tourist outlets and bookshops throughout the country. He was frequently sought out by visiting researchers because of his expertise, which was collected over many years of making sampling runs in the region, and despite the fact that he has no formal qualifications in the sciences. During collecting trips he had both discovered entirely new species and found species growing in the Monteverde region that were previously not known to flourish there.

---

[9] Monteverde's endemic orchid species are epiphytes, and so grow naturally by attaching themselves to trees or other plant life for better access to sunlight and increased opportunities for seed dispersal (see Forsyth and Miyata 1984: 42).

Despite his obvious enthusiasm and the energy he devoted to both his research and his business, Rafael echoed the sentiments of many other residents – conservationists, scientists, educators, and business owners alike – who worried that people in the community thought only about making a profit, and not about the environmental impacts of what they did:

> People who live here may know the facts about environmental issues because they have been taught them, but that doesn't mean that they have developed the necessary conscience to stop polluting and damaging. But how can you teach that conscience?... The irony is that everyone leaves Monteverde alone because it has this great reputation for environmental success, but there are so many problems remaining here. People in the community protect the trees because that is what brings them economic benefits. They will only start doing other things to protect the environment if they can see a similar way of profiting from it.

In addition to many expressions of concern about the influence of the growing local tourism industry and its impacts on the local environment and quality of life, however, Monteverde residents and business owners also spoke about ways in which ecotourism and educational tourism had already brought positive change to the community. Carlos Soto, co-owner of the local frog exhibition – known as the *Ranario* – who was born and raised in the region, commented:

> 'Maybe 15 years ago, people here thought that tourists were a little crazy for travelling so far to see a tree or a monkey. They were happy to show them, of course, and to make money off the tourists who wanted to take pictures, but no one understood what was so exciting about a monkey in a tree… Now things are really different. People are proud of their national patrimony and they can see now that its value is greater than just the money that can be made from it.'

One important reason for this change, he believed, was due to the economic benefits that tourism has brought to the region: 'People could begin to think about things other than just survival. We could begin to learn more and educate ourselves'. According to Carlos, when his grandfather came to settle in the region in the 1930s, there was not yet any tourism infrastructure, and even as late as 1981 there was only one hotel in the village of Monteverde: 'In those days people used to come here looking for the beach. As far as tourists were concerned, Puntarenas was the beach, and they expected to find it here… Some even brought surf boards', he told me, laughing.

Carlos got his own start in ecotourism by working as a nature guide in the Monteverde Reserve from 1995 until 2000, and it was largely as a result of that experience that he wanted to open a business related to conservation and education. He decided to work with endemic species of frogs and toads because while working as a guide he realised that many people were afraid of them:

> 'The reason they are afraid is because they don't know much about them. When they visit the Ranario, little by little they become less afraid and more informed. And when students visit they can see the possibilities for future studies and even begin to imagine themselves as scientists and researchers themselves.'

In all of our conversations, Carlos emphasised that creating and maintaining good relationships with the community were an important part of his business:

> 'I would like for as many people as possible to be able to visit. I'm here every day, so I know the people from the area and I am always gratified to see them come again and again.

There is one grandmother I know who brings her four grandchildren almost every Saturday, because the kids just love it.'

All local residents were welcome to visit the exhibition for free at any time, and support was provided for student groups from further afield who wanted to visit but were short of funds. This took the form of reduced entry charges or help with transportation costs. In addition, the organisation operated a 'neighbours' (*vecinos*) scheme which extended from the community of Guacimal to the town of Las Juntas [distances of approximately 35 km south and 60 km northwest of Santa Elena along the main road, respectively], and which continued to be actively expanded. Residents of these designated 'neighbour' communities were eligible to enter the exhibit for free at any time.

While there was broad and strong support for the project from many diverse community members in 2003, however, this had not always been the case. When Carlos and his business partners – one of whom was elected mayor in 2002 – first began to develop the business in 2000, they met with serious resistance from scientific researchers working in the region. What began as a largely personal dispute between Carlos and a family member over the business itself and the rights to its future profits, quickly escalated into a prolonged legal battle with officials in the Ministry of Environment and Energy (MINAE) and even a stand-off with armed police. On either side of the conflict were local residents – scientists, conservationists, business owners and family members – with opposing viewpoints about who in the community possessed the relevant expertise to safely care for the sensitive amphibians in captivity, and also to pass on knowledge about them to community members and visitors. Such conflicts highlighted the long-standing tensions in the community regarding access to local environmental knowledge and, specifically, about who possessed the relevant knowledge to effectively manage the protection and study of local resources.

Negotiating Environmental Knowledge in the Community

All of these perspectives and stories of ecotourism development clearly highlight the ways in which negotiations over access to local resources in Monteverde were not just limited to the practical management of local flora and fauna. Access to knowledge about these natural resources and the accompanying ability to use that knowledge in particular ways – for example, by providing educational experiences for residents and visitors or using it as the basis for a profitable local enterprise – were also important for local scientists, conservationists, and business owners. Environmental knowledge, in this sense, was a kind of commodity to which diverse community members negotiated access, and environmental education of various types (in schools, by conservation NGOs, in public debate) was a key site for this active negotiation.

Participation by individual residents in public spaces organised by local organisations and interest groups was therefore an integral part of processes of knowledge formation and dissemination in the community. Parents who took their children to visit local educational tourism destinations, for example, were involved in both implicit and explicit decisions about the ability of particular local ecotourism business owners to promote and provide 'appropriate' or 'useful' knowledge and information about the local environment. Similarly, residents who took part in protests related to the establishment of the *Ranario* were also actively engaged in these debates.

Perhaps most interestingly (and as already indicated in Chap. 3), much local debate and discussion in 2003 actively challenged the historical dominance of foreign scientists in local conservation and argued for other members of the community to have a much greater role in conservation, local development and education. Many business owners, however, told me that they remained frustrated both by the limits to existing research and local access to it. Carlos Soto, in particular, noted that the amount of research examining the behaviour of amphibian species in captivity – information which he argued was important for both his business and to advance scientific understanding generally – was exceedingly limited. He identified the source of this problem as the result of disagreements between scientists, conservationists and ecotourism interests:

> 'The people that study frogs here [in Costa Rica] are a very closed group. They don't share information… in fact, there really isn't information in Costa Rica on frogs in captivity because the people that study them don't believe in keeping them in captivity… There are symposiums all over Latin America where other people talk about how we should work towards reproduction in captivity so that populations can be re-built, but the scientists don't believe in that.'

Similar concerns about research and access to results on many other environmental topics were highlighted during discussions that I had with educators, business owners and staff of local development organisations in Monteverde throughout the year. Particularly strong critique from many individuals was aimed at the dominance of the protectionist conservation agenda – actively promoted by many local conservation groups and individual researchers – which has characterised conservation efforts in the region since the 1970s.

This protectionist style of conservation has been very successful in terms of marking out vast territories in the region that are entitled to legal and administrative protection from either local organisations or the state. Alongside the credit given to local organisations and individual scientists for this success, however, there was also considerable resentment from some community members – including educators, staff in local development organisations and business owners – regarding the relative lack of access to research results and data collected about local ecosystems. Several initiatives were begun by local groups in 2003 to promote wider access to this body of information. Proposed projects included the establishment of a community information centre and the increasing use of public consultations, but access to research results remained a point of significant contention. The vast majority of available reports and papers, for instance, are presented only within academic publications (which are often difficult to access) and in the native language of researchers (usually English).

This lack of communication had concrete effects throughout the community, and especially for local education, as I observed in August 2003 when I was asked to help judge the *Colegio*'s annual science fair. On the morning of the fair, I arrived at the school to find the students' displays set up in a classroom. In total, there were 18 entries of varying categories, dealing with topics ranging from methods for making natural dyes using local flora to research projects on endemic birds, snakes, frogs, and butterflies. When questioned about the sources for their information, the student research groups recounted visits to local ecotourism sites and assistance from locally resident scientists. One young woman presenting a group project on endemic birds, for example, told us that her older brother (employed by a local conservation organisation) had helped her get access to rare video footage made by a local researcher of a particular mating dance. Many other students, however, complained about a lack of access to publications or information, and even those with connections to individual researchers or local organisations through their own family or social links protested bitterly that the majority of the resources they found were available only in English. Over lunch later that day, a fellow judge, a young Costa Rican woman employed by a local research organisation, also expressed her personal frustrations about relationships between researchers and the wider community. As one example of this, she cited the case of Rafael Carazo, owner of the orchid garden, whom she told me was routinely consulted for his expertise in local flora by visiting scientists, but rarely received compensation or credit for his contributions.

On yet another occasion, I asked an environmental educator in Santa Elena to tell me about the links between the local scientific community and local environmental education. He replied:

> *In my opinion, there isn't much communication at all. The Tropical Science Center, in particular, guards their information and it's not available to the general public... It all goes straight to San José. As for the Institute, the information is there, but either it's in English or it's not of interest to anyone here. The Monteverde Conservation League only does a little research, so they aren't much help either. Mostly the scientific research is being conducted by independent researchers who may or may not be affiliated with anyone local. Lots of them work without even getting approval from the Ministry of Environment, so whatever data they uncover just goes straight back to wherever they came from.*

These frustrations over access to information were, of course, partially practical – in that educators and other local residents expressed a desire to use the information for their own purposes. However, I believe that they were also connected to larger, underlying struggles regarding who in the community was 'qualified' to pass on knowledge about the local environment, and to make decisions about the kinds of education promoted in schools and in the wider community. While locally-resident US conservationists tended to argue that scientific or academic training and expertise were the most important factors in taking on this role, many Costa Rican business owners – like Carlos Soto and Rafael Carazo – believed that their years of residence in the community, as well as their independent study and experience of these topics made them equally qualified to engage with visitors, students and other community members.

The links between local institutions – including schools, local conservation groups, municipal officials and other kinds of NGOs – in terms of environmental education programmes should therefore be seen as an indication of the ways in which diverse kinds of environmental knowledge and perspectives on education circulated through community networks in Monteverde. These networks in turn were embedded in local economic, social and political relationships which impacted upon how residents participated (or did not) in particular initiatives. When groups received significant support from the community, they were able to promote particular perspectives on environmental knowledge in accordance with particular perspectives and goals, and in turn to influence the character and content of local environmental learning. The process of environmental education, in this understanding, is therefore not simply the neutral provision of information in institutional spaces (such as schools or protected areas), but rather is a complicated interaction between diverse forms of knowledge and learning, and powerful community relationships.

Conclusions

The accounts of meetings and events in public spaces outlined in this chapter detail the dynamic negotiations involved in the implementation of environmental education within the community, as well as the institutional logics and relationships shaping them. In Monteverde, these negotiations were particularly strongly linked to struggles over access to local resources – including both natural resources themselves as well as knowledge about them – which were variously used by local residents to support local environmental management, to promote particular community development agendas, and to build profitable businesses and professional careers. These struggles were in turn connected to the many different understandings of 'development' expressed by individuals and organisations located in the community. This diversity of perspectives resulted in moments of both collaboration and competition, and the ways in which local residents with diverse interests participated (or not) in the work of local organisations thus gave support to the promotion of particular kinds of educational and community development programmes.

These accounts of public spaces also serve to illustrate the importance of extending an analysis of environmental education beyond activities within particular institutional or organisational settings in order to account for the ways in which environmental knowledge and educational practices are embedded in complicated economic and social relationships. It is, of course, important to understand how local schools manage the limits on environmental learning imposed by the state education infrastructure, and how local conservation groups negotiate programming content and goals within particular frameworks of organisational commitments and community relationships. Added to this, however, should also be an attention to other kinds of sites – and particularly those such as public spaces that are often not included within discussions of educational processes – in which particular kinds of environmental knowledge are actively promoted and negotiated through complex networks of social, economic, and political relationships.

From this perspective, it is possible to explore the importance of these sites in Monteverde as reflexive spaces, and also to examine the ways in which public debate and discussions informed and influenced public opinion, impacted upon the practical implementation of community projects, and fed back into ideas about the 'appropriate' content and pedagogical orientations of more formal environmental education programmes such as those sponsored by local schools and conservation organisations. Environmental education within the community can in this way be seen to be deeply embedded within a dynamic 'educational infrastructure' composed of particular institutions and sites for learning, the economic, social and political relationships between local residents, and active negotiations and contestations of knowledge. Environmental education is therefore not simply about what kind of information (science/social concerns) is passed on to whom (students in schools, adults in protected areas) or in what ways (transformative or skills-based pedagogies), it is also about much more deeply rooted negotiations over knowledge and resources within a community.

References

Anderson, J. R., Reder, L. M., & Simon, H. A. (1996). Situated learning and education. *Educational Researcher, 25*(4), 5–11.

Asociación Agua Pura. (2003). *Informativo Agua Pura*. Volume 5, Number 19. Community newsletter produced by Asociación Agua Pura and the Asociación Administradora del Acueducto y Alcantarillado Sanitario del Cantón de Monteverde. Monteverde, Costa Rica.

Bandura, A. (1977). *Social learning theory*. Englewood Cliffs: Prentice Hall.

Burlingame, L. (2000). Conservation in the Monteverde Zone: Contributions of conservation organisations. In N. Nadkarni & N. Wheelwright (Eds.), *Monteverde: Ecology and conservation of a tropical cloud forest* (pp. 351–388). Oxford: Oxford University Press.

Dallas, S., Scheffe, B., & Ho, G. (2001). *A community-based ecological greywater treatment system in Santa Elena-Monteverde, Costa Rica*. Paper for the IWA Conference: Water & Wastewater Management for Developing Countries, Kuala Lumpur.

Forsyth, A., & Miyata, K. (1984). *Tropical nature: Life and death in the rain forests of Central and South America*. London: Simon & Schuster.

Glasser, H. (2007). Minding the gap: The role of social learning in linking our stated desire for a more sustainable world to our everyday actions and policies. In A. Wals (Ed.), *Social learning towards a sustainable world: Principles, perspectives and praxis* (pp. 35–61). Wageningen: Wageningen Academic Publishers.

Hager, P., & Hodkinson, P. (2009). Moving beyond the metaphor of transfer of learning. *British Educational Research Journal, 35*(4), 619–638.

Hanks, W. F. (1991). Forward. In J. Lave & E. Wenger (Eds.), *Situated learning: Legitimate peripheral participation* (pp. 13–24). Cambridge: Cambridge University Press.

Honey, M. (1999). *Ecotourism and sustainable development: Who owns paradise?* Washington, DC: Island Press.

Illeris, K. (2007). *How we learn: Learning and non-learning in school and beyond*. London: Routledge.

Keen, M., Brown, V., & Dyball, R. (Eds.). (2005). *Social learning in environmental management: Towards a sustainable future*. London: Earthscan.

Lave, J., & Wenger, E. (1991). *Situated learning: Legitimate peripheral participation*. Cambridge: Cambridge University Press.

Leeuwis, C., & Pyburn, R. (Eds.). (2002). *Wheelbarrows full of frogs: Social learning in rural resource management*. Assen: Koninklijke van Gorcum.

Maroto, L.R. (1997). *Evaluación de Monteverde 2020 y propuesta metodologica*. Unpublished doctoral thesis, Universidad Nacional, Heredia, Costa Rica.

MVI [Monteverde Institute]. (2002a). *Encuesta de desarrollo: encuesta para residencias*. Unpublished survey data from the Monteverde Institute and the USF Globalization Research Center. Monteverde, Costa Rica: Monteverde Institute.

MVI. (2002b). *Encuesta de desarrollo: comercio y ecoturismo*. Unpublished survey data from the Monteverde Institute and the USF Globalization Research Center,Monteverde, Costa Rica: Monteverde Institute.

MVI. (2003). *Final report: A study of water management and health in the household in three communities in the Guacimal River watershed using rapid assessment procedures* (Unpublished Research Report No.). Monteverde, Costa Rica: Monteverde Institute.

Salomon, G., & Perkins, D. (1998). Individual and social aspects of learning. *Review of Research in Education, 23*, 1–24.

Wals, A. (Ed.). (2007). *Social learning towards a sustainable world: Principles, perspectives and praxis*. Wageningen: Wageningen Academic Publishers.

Wenger, E. (1998). *Communities of practice: Learning, meaning and identity*. Cambridge: Cambridge University Press.

Chapter 6
Conclusions

Abstract Environmental education has been at the centre of international and national policies of sustainable development for the last several decades, and has stimulated significant debate regarding both its inclusion in curricula and proposed methods for implementation. The research on which this book is based used anthropological fieldwork to explore environmental education and learning with schools and non-governmental organisations and in public education spaces in Monteverde, Costa Rica. This chapter revisits the main arguments of the book and suggests some potentially useful ways forward for future research.

Keywords Community development • Environmental education • Learning • Sustainable development

As the preceding chapters have shown, in 2003 environmental education in Monteverde was a key site for the contestation of understandings of the natural world and humans' relationships to it, as well as sometimes a catalyst for conflicts over access to natural resources and knowledge about them. Diverse individuals and organisations in the community were actively engaged in these struggles during the time of this research, and these engagements were mediated by economic and social relationships both within the community and across local, national and international contexts.

The powerful influence of locally-resident US scientists and conservationists has meant that scientific discourses of environmental protection have historically dominated policy and practice in the Monteverde region. This dominant way of knowing and managing the local environment – largely through the establishment of strictly protected forested areas – in turn influenced the content and orientation of many local educational programmes. Through education both in formal school programmes and in wider community education efforts by local organisations, the perspectives of residents with an interest in such styles of environmental management also fed strongly into community development agendas.

Alternative perspectives on the environment and local development, however, were strongly advocated by those who fundamentally disagreed with this protectionist approach for a variety of reasons. This included, for instance, business owners who sought access to local natural resources in order to run profitable enterprises as well as participants in community development organisations which promoted a style of development that aimed to improve local infrastructure and services. Resistance to prevailing styles of local environmental management had taken many forms, including outright public protest and the establishment of new conservation groups, businesses and business interest groups which organised educational efforts and engaged in public discussion, as well as complaints from (mostly Spanish-speaking) students, teachers and staff in local organisations over limited access to research findings. These kinds of activities worked against dominant public representations of local environmental concerns, as well as the often taken-for-granted notion that these could be most effectively managed through strict protection by local conservation organisations. In these ways, public debate and educational programming were part of active local critique of both dominant (scientific) ways of knowing and managing natural resources, and of the perceived economic and social inequalities which tended to accompany them.

Environmental education and learning in the community were therefore deeply entangled in local negotiations of community development and environmental management, and were significant touch points for both collaboration and conflict. Such conflicts revealed deep fault lines in the community, often rooted in differences in language, nationality and educational background, or social and family ties. Residents involved in education, environmental management and community development were not only keenly aware of these fault lines, but routinely sought to work both within the limits they imposed and across them. At an organisational and an individual level, local residents were often situated in a number of complicated ways, and made claims to membership in, or affiliation with, particular kinds of influential networks that facilitated access to available resources. Costa Rican business owners in Monteverde, for instance, had very strong links with national policy makers, business interest groups and trade associations. These affiliations were often established through the course of business ventures or were rooted in strong networks of family and social acquaintance. Locally-resident foreign scientists, on the other hand, could call upon networks of international colleagues, including contacts in foreign universities and research institutions, as well as international conservation or policy organisations, in order to gain access to funding or support for projects or initiatives. In both cases, these influential individuals and groups could mobilise their networks in order to impose pressure on local or national policy makers about particular issues or concerns.

These struggles over knowledge and resources were deeply interwoven with key debates about the appropriate content and goals of environmental education in the community. As highlighted throughout the book, two competing narratives of environmental education could found both in Monteverde during the time of this research as well as within theoretical debates in the research literature. Advocates of the first approach – usually scientific researchers and conservationists – tend to privilege the promotion of scientific understandings of the natural world, and emphasise that increasing public knowledge of this kind will result in changes to damaging or unsustainable behaviours (a commonly-heard phrase is that 'if people learn to love the natural world, they will work to protect it'). Advocates of the second view, on the other hand – often educators and community activists – see environmental education as a process which promotes critical thinking and engagement, and through which learners are encouraged to understand and respond to the interactions between the natural environment and human communities.

The complicated nature of local identities and relationships in Monteverde means that it would be far too simplistic, however, to draw lines of division between groups or perspectives based solely on language (Spanish or English), nationality (Costa Rican or foreign), or profession (conservationists or tourism operators). Significant, and continuing, inward migration from other parts of Costa Rica and the rest of the world since the 1950s, as well as long-term settlement by both Quakers from the US and scientific researchers (many of whose children now claim both 'Tico' and 'gringo' identities) – has substantially blurred the lines between 'insider' and 'outsider' status in the community. Equally, while many local tourism business operators actively sought to join networks of local and international conservationists and to engage in the promotion of strict local environmental protection or of local 'sustainable development' initiatives, others aligned themselves more closely with local development associations and emphasised the promotion of 'development' through the modernisation of roads, transport and communications. These diverse engagements highlight the ways in which individuals and organisations in the community negotiated multiple, and sometimes conflicting, identities and relationships as part of wider negotiations of environmental knowledge and power in the local context.

The community was an excellent site for this research largely because of this diverse population of individuals and organisations actively engaged in education, conservation and community development, and also due to its long history of strong national and international connections. The diversity of perspectives represented by community members and the disparate types of, and sites for, their engagement with environmental knowledge collectively constituted a complex network of interactions within the community which were in turn embedded in multiple social, economic and political relationships. These networks and relationships cut across local, national and international contexts and had concrete impacts on processes of education, environmental management and development.

At the beginning of this book, for example, I offered an outline of the history of education and educational ideologies in Costa Rica, and highlighted some of the reasons why environmental education has proved to be so popular with both policy makers and the public (Chap. 2). These included a long-term national emphasis on citizens' entitlement to education, the discursive importance given to the role of education in national social and economic development, and the state's strategic focus on the promotion of scientific research, knowledge and conservation. Despite the strength of discourses about the importance of education, however, the Costa Rican state has often been unable to effectively implement national education policy, and especially programmes promoting environmental learning. Problems with the state system that were frequently cited by state educators included the massive and inefficient nature of the state education bureaucracy, as well as a long-term lack of sufficient infrastructure and resources. The state has, however, managed to attract much-needed aid and investment from other nations and international organisations, largely through its heavy promotion of education, conservation and research. Such international support for conservation and education has both mediated the impacts of the state's financial difficulties and brought with it new, and sometimes conflicting, ideas about the appropriate content and provision of educational programmes.

The complicated nature of these interactions between individuals and organisations with diverse perspectives on environmental education were particularly well-illustrated at the local level in Monteverde during the time of this research, where educators in a variety of organisational settings and circumstances negotiated its implementation in practice. Local schools, for instance, acted as important sites for knowledge transmission, negotiation of meanings and priorities, and processes of economic and social development (Chap. 3). Despite their many differences – in terms of funding levels and sources, curriculum content and pedagogical orientation – all of Monteverde's schools were tied to strong relationships with the state, as well as to local social and economic relationships which impacted upon educators' decisions regarding implementation of the curriculum and approaches to classroom practice. These included limitations imposed by the state education bureaucracy (and especially the demands of the national assessment system), the demands of parents and local employers, and on-going problems as a result of insufficient teaching and financial resources.

Such severe limits on resources led many schools in Monteverde to rely heavily on educators employed by local conservation organisations to provide environmental education programmes. Educators working for these conservation groups, however, also faced significant challenges to programme implementation as they managed complicated relationships with diverse community members, and between local, national and international interests (Chap. 4). Using detailed case studies of the environmental education co-ordinators at two local reserves, I highlighted the ways in which their very different environmental education programmes were both constrained and supported by individual commitments to environmental education, by the goals and agendas of their respective organisations, and also by the relationships of each organisation to the wider community and to national and international partners.

Local conservation organisations were not the only local groups to be so heavily enmeshed in relationships to wider local, national and international contexts. Indeed, the majority of local organisations made decisions about policies, projects and programmes in the midst of on-going tensions regarding local environmental management and community development (Chap. 5). Historically, many local organisations and individuals, and particularly scientific researchers, have been committed to promoting a strictly protectionist conservation agenda. During the time of this research, however, local debates had begun to centre more heavily on the need to find a balance between protection of local forests, promotion of the local tourism industry, and improvements to local infrastructure. Local residents used public spaces and events – such as educational tourism destinations, meetings and workshops – to learn about local concerns, and to debate and discuss solutions. Public spaces were thus important to community members both as sites for gaining access to information about local concerns and also as sites for actively debating community development decisions.

These varied accounts of environmental education in state and private schools, from local conservation groups, and within other public spaces in Monteverde illustrate the dynamic negotiations of environmental knowledge occurring within the community, as well as in relationships with a wide variety of national and international actors and organisations. They also serve as a clear example of the reasons why an analysis of environmental education must look beyond programmes in individual settings in order to account for the ways in which environmental knowledge is situated within complicated networks of economic and social interaction. This book has therefore attempted to capture the complex and dynamic processes surrounding the dissemination and creation of environmental knowledge in multiple sites and across multiple levels, and has also highlighted the need to view these processes as influenced by both ideological and pragmatic concerns.

While much of the existing policy at both international and national levels continues to imply that implementation of successful environmental education is largely a matter of allocating sufficient funding and effectively targeting local populations with the 'appropriate' style of education, this research also suggests that such an approach rests on a rather simplistic understanding of the nature of learning. In doing so, it neglects to give attention to the economic, social and political fault lines which run across local and national landscapes of environmental education practice. The book further argues that the complexities of these interactions and the many levels on which they occur requires a more expansive approach to research and attention to multiple sites of engagement with education and learning.

Contributions to Existing Research

In much broader terms, the research set out to explore the ways in which understandings and practices of environmental education are shaped in a particular context. This kind of analysis is necessary because there has so far been relatively little active research engagement with issues surrounding the implementation of international policies of environmental education 'on the ground' or of the perspectives that underpin them. Initiatives such as *Agenda 21* and the UN Decade of Education for Sustainable Development, for example, rest upon a common assumption that increasing public knowledge of environmental issues is an inherent good, but the potential challenges and concerns that surround implementation of programmes in particular nations or communities are not yet fully understood.

Many practitioners and academics also continue to debate exactly what environmental education is and how it should be implemented (cf. McKeown and Hopkins 2003; Kollmuss and Agyeman 2002; Dillon and Teamey 2002; Stables 2001; Palmer 1998; Huckle and Sterling 1996; Jickling 1992). Despite a perhaps universally shared goal of stimulating social change through education and awareness-raising, a wide variety of definitions and perspectives on environment and education have resulted from these debates – including not only theories of environmental education, but also of 'education for sustainable development', development education, 'education for sustainability' and 'education for a sustainable future', among others. These debates about terminology and conceptualisations in turn raise some fundamental questions: Is the aim of these efforts to teach particular sets of facts (for example, taxonomies and scientific understandings of the natural world) or to awaken learners' consciousness of environmental problems and their responsibility to help ameliorate them? How might these lessons be most effectively taught – through classroom lectures, learning experiences in forested areas, or as part of group discussions about environmental ethics and social inequality? And, perhaps even more importantly, how are these concepts and approaches understood in diverse national and local contexts?

The recent emergence in some contexts of the term 'climate change education' has further energised these discussions (cf. Læssøe et al. 2009). As of yet, the term appears to be mostly part of efforts to raise the public profile of climate change as a single issue, and the concept has yet to be explored in any depth by academic research. As a result, it is probably too soon to know how the concept might develop, but its emergence raises a number of issues that are familiar in related fields. For instance, should the central role of climate change education be to teach people (of all ages) particular 'facts' about the state of the world's climate and as a result to encourage them to perform certain pre-determined, 'correct' behaviours (e.g. conserving energy, recycling, reducing carbon consumption)? Or should it be to more broadly to support them to develop the capacities to address rapid environmental and social change and future uncertainty (e.g. through critical thinking skills and understandings of global inter-relationships)?

As I outlined in the introduction to the book, underpinning these questions are fundamentally different perspectives on the role of education in individual, social and environmental change. While a broad spectrum of understandings are represented both in research and in practice, two rather distinct perspectives also seem to emerge. The first sees education largely as a process of knowledge transmission and aims to promote particular kinds of attitudinal or behavioural change (e.g. to more 'environmentally responsible' behaviours). The second argues that education is a process which encourages learners to develop skills, such as critical thinking and problem solving, as well as to encourage flexibility and adaptability, in order to enable them to address the challenges of sustainable development and climate change, and of living in a rapidly changing world more generally.

Given the complex and rapidly changing era of globalisation in which we now live, these concerns continue to take on ever greater significance. What kind of environmental learning will be needed in order to effectively cope with the global social, economic and environmental changes we are likely to encounter in the future? Where should this learning take place – in formal education, through community initiatives, in public spaces or in multiple locations? And, especially in the context of the current global economic crisis, how might environmental learning fit – or not – within existing agendas for educational development both in 'developing' and 'developed' country contexts?

In Monteverde in 2003, a large number of individuals and organisations were engaged in teaching and learning of various types and in a wide range of educational settings – including, but not limited to, schools, initiatives within local protected areas, and public spaces. While the many groups and individuals involved in these activities did not always agree on the best means by which to deal with local environment and development issues – and so would likely have answered the questions raised above in rather different ways – their sometimes conflicting perspectives and practices nevertheless formed a strong network of discussion, effort, and innovation across the community. In turn, this offered community members with multiple opportunities to both learn about environmental topics or other issues of local concern, as well as to address local concerns occurring in the community *outside* these sites, and especially related to the character and progress of local development and environmental management. This suggests that promoting such debate and discussion can be highly productive at the community level, both in terms of encouraging critical exploration of environmental and sustainability issues, as well as more generally promoting public awareness of and engagement with those concerns.

Environmental Learning in Latin America

This research was also intended to serve as a useful starting point for conducting similar research on environmental education in other parts of the world. In particular, it adds to both existing work on environmental education (cf. Pellegrini Blanco 2002; González Gaudiano 1999) and on the ethnography of education (cf. Levinson et al. 2002; Anderson and Montero-Sieburth 1998) in Latin America. Further research in other sites in the region would contribute to a much deeper understanding of how negotiations of environmental knowledge and educational programming are both similar and different, for instance, in areas of post-conflict reconstruction (for example, in the case of emerging environmental education programmes in Guatemala and El Salvador) or in areas characterised by the strong influence of indigenous politics (for instance, educational programmes linking modes of environmental management with local indigenous cosmologies or self-determination efforts both within Costa Rica and elsewhere in the region).

As I have argued throughout the book, practices and perspectives of environmental education are deeply embedded in particular social, historical and economic contexts, and so it is important for any such comparative work to acknowledge that interpretations of what constitutes 'appropriate' programme content or teaching methods are deeply connected to the specific contexts in which they take place. Certainly, the cases of both Monteverde and of Costa Rica – with their reputations for peaceful democracy and successful environmental protection – may be considered somewhat unusual in comparison to many Latin American neighbours. Nevertheless, the research highlighted a number of shared characteristics and concerns with other countries in the region, including the challenges of transversal/interdisciplinary teaching and learning approaches and the complications inherent in historical and contemporary global economic, political and social relationships.

Firstly, the inclusion of environmental learning as a 'transversal' theme within the curriculum is a common feature in many Latin American education systems, likely due to its roots within an earlier Spanish educational reform movement (González Gaudiano 2007). As in Costa Rica, these themes are intended to cross-cut all other areas of the curriculum and provide multiple opportunities for engagement with particular topics through integrated, cross-disciplinary activities (cf. García Gómez 2000; Roth 2000; Lencastre 2000; Luzzi 2000; González Gaudiano 2000; Reigota 2000). While there has been relatively little analysis or evaluation of these efforts so far in the region, the existing research suggests that educators often find transversal themes difficult to manage due to a lack of appropriate training, support and resources, as well as due to their wider conceptual and pedagogical challenges. These concerns mirror those expressed by a number of educators both in Monteverde and in the national Ministry of Education offices who noted the potential value of the themes as part of the curriculum, but found their implementation difficult in practice due to both a lack of training and resources, as well as the demands of the national examination system.

Discussions about transversal themes in Latin America also resonate with wider international conversations about the use of interdisciplinary teaching and learning as tools for addressing sustainable development (cf. Læssøe et al. 2009). In the UK, for instance, the previous government's National Framework for Sustainable Schools identified the need for 'whole school' approaches to sustainability, and introduced teachers and schools to eight 'doorways' through which they could initiate activities: food & drink, energy & water, travel & traffic, purchasing & waste, buildings & grounds, inclusion & participation, local well-being, and participation (DfES 2006).[1] UK teachers, however, often express uncertainty about how to apply these principles to their everyday practice, perhaps because few teacher education programmes include strategies for teaching about sustainability or for addressing interdisciplinary topics more generally. A recent report by Ofsted (the body responsible for school inspective in England) also found that limited provision within individual subject areas – including citizenship, geography, science, and design and technology – resulted in few opportunities for the kind of cross-curricular learning which such government policy has advocated, although there was more evidence of this kind of learning in primary schools where planning more easily crosses subject boundaries (Ofsted 2008).

Secondly, the diversity of perspectives on and approaches to environmental education that are found in Monteverde reflects a similar spectrum of ideas within Latin America more broadly. This is likely due to the wide variety of pedagogical and political traditions (including liberation theology, dependency theories, popular education, adult education) and thinkers (e.g. Paulo Friere, José Carlos Mariátegui,

---

[1] It is uncertain at this point whether the new UK government, which took office in 2010, will either alter, continue, or discontinue this policy initiative.

José Martí, Simón Rodriguez) to which environmental education has historically been linked in the region:

> It thus varies from socially accepted compensatory programmes designed to help academic strugglers such as illiterates and school drop-outs and social integration programmes (i.e. teaching Spanish to monolingual indigenous populations, street children), to others of a more libertarian nature that upset the existing social order (i.e. *guerrillas* and insurrections against despotic caciquism). (González Gaudiano 2007: 159)

The ways in which environmental education in the region has been tied to this range of broader social concerns has therefore resulted in a stronger political and activist orientation to the field than is typically found in Europe or North America.

This political orientation stands somewhat in contrast, however, to conservation and education initiatives by many international scientific and conservation organisations working in the region. Despite the endorsement of the term 'education for sustainable development' by many of these influential organisations (including, for instance, UNESCO and the International Union for the Conservation of Nature and Natural Resources), these efforts nevertheless tend to focus on the protection of areas of global scientific importance (often through land purchase campaigns), the continuing development of scientific understandings of tropical biology, ecology and forestry, and the promotion of conservation education/environmental education programmes which support those goals.

While such support by international organisations is often clearly welcomed by policy makers, conservationists and educators, understandable questions are also asked about the fairness of asking 'developing' nations to protect their natural resources (and to forgo particular forms of industrial development) for the benefit of the rest of the world. This is perhaps most apparent with regard to the issue of climate change, for which there is an obvious inverse relationship between historical responsibility, which predominantly lies with industrialised countries (research shows that CO_2 levels began their dramatic rise following the industrial revolution), and those countries that are most vulnerable to its impacts. These tensions were clearly highlighted at the UN climate change summit in Cancún in 2010, for instance, where developing country representatives argued that industrialised nations have the major responsibility to reduce CO_2 emissions and to address the root causes of climate change.

Moving Forward: Areas for Future Research and Policy in Environmental Education

All of the complex relationships highlighted in this research, as well as within such contemporary international policy and discussions, bring us full circle back to the central arguments of this book.

Firstly, there is a need to expand the focus of research on environmental education (and related areas) beyond single educational sites such as classrooms, schools or protected areas. While this focused research is clearly important for the field, there is also a need for greater exploration of the relationships *between* different educational sites and *between* theory and practice. As the case of Monteverde shows, activities in these sites do not happen in isolation from the wider communities in which they are located. Instead, they are strongly linked to the historical, political, economic and social relationships that characterise the local context, as well as to broader national and international contexts. This is in line with the growing acknowledgement within research of the socially-embedded nature of environmental education – including the recent exploration of 'free choice' learning (cf. Falk 2005) and of the social dimensions of learning (cf. Wals 2007; Reid et al. 2008) – and of the need for concepts and methodologies to adequately explore it.

Anthropologists have also long taken an active interest in education, and particularly in its links to wider social life (cf. Wax et al. 1971; Spindler 1963). More recent research has begun to even further extend the borders of anthropological engagements with education by examining the ways in which individuals use social spaces to contest public representations and to critique existing structures of power (cf. Boyte and Evans 1992; Fine and Weis 1998; Fine et al. 2000). Although research in the anthropology of education has yet to give much attention to environmental education as a topic of research interest, this shift in the research agenda signals an important movement away from narrow attention to educational and learning processes in single sites, and towards a fuller exploration of the links *between* these sites and of the ways in which individuals learn, negotiate complex networks of social, economic, and political relationships, and also actively engage in the promotion of particular kinds of knowledge. In this way, the book also dovetails with existing work in the anthropology of development (cf. Grillo and Stirrat 1997; Gardner and Lewis 1996) which seeks to identify not only the overtly powerful influence of particular groups or individuals in the development process, but also the multi-levelled and multi-sited negotiations of meaning and practice that are a part of wider processes of both change and resistance.

Secondly, there is a real need for research which explores environmental learning as a process and learners as active participants. As several authors have already suggested, more work is needed to more fully understand the complex learning processes that take place when individuals engage with environmental and sustainability topics (cf. Rickinson 2001; Scott and Gough 2003; Heimlich and Ardoin 2008; Rickinson et al. 2009). Perspectives and stories from Monteverde illustrate that local residents were not simply passive recipients of educational messages, but rather were actively engaged in negotiating knowledge about local environmental and development issues.

Of course, no research project is perfect, and there were several issues that – for reasons of time constraints or lack of access – I was not able to explore in any depth. For instance, in Costa Rica and around the world, both government agencies and NGOs of various kinds (local, national and international) are increasingly using online spaces to disseminate environmental messages and to encourage participation in environmental management and sustainable development. International school partnerships, many of them facilitated via email and online discussion boards, are also proving increasingly popular in the UK and Europe. All of these new opportunities pose a range of new challenges for environmental teaching and learning, including the need to ensure that online resources support good quality learning, to account for and sensitively manage diverse perspectives on education and environmental concerns, and to address the inherent imbalance of power between educational institutions and organisations located in 'developed' and 'developing' country contexts. Research which explores these aspects of new communications and learning – either directly or indirectly related to environmental education – is needed to address these challenges for both policy and practice.

Further work is also needed to explore the relationships between environmental education and dimensions of social inequality such as gender, ethnicity and socio-economic status. Although these have been key areas of concern for mainstream educational research for a very long time, they have yet to be strongly taken up with regard to their impacts on environmental learning specifically. National discourses in Costa Rica, for example, are framed in a vocabulary of empowerment and participation, and draw heavily upon historical accounts of national social and economic development as egalitarian and democratic. In daily life, however, real concerns about social and economic inequality, and the ways that they impact upon individual educational achievement and national social and economic development – and in turn on environmental management and sustainable development efforts – are often marginalized or entirely neglected.[2]

Previous research by anthropologists (cf. Stocker 2005; Leitinger 1997) and educational researchers (cf. Palmer and Rojas Chaves 1998; Twombly 1998; Stromquist 1992), as well as the research conducted for this book, have also highlighted some of the ways in which unequal gender relations in the country, in particular, have serious implications for education provision in Costa Rica. These impacts are experienced in relationships between and among students, teachers, and policy-makers on an individual level, as well as within classrooms, schools and the Ministry of Education. The education sector in Costa Rica, as in many other nations, is overwhelmingly staffed by women – they fill the vast majority of classroom teaching posts around the country (especially in primary schools) and make up a significant proportion of the national educational bureaucracy (although often at the lower levels of the hierarchy). Research regarding these concerns in any context requires an exploration both of national policy discourses, as well as of practical, day-to-day experiences of the limitations and opportunities encountered by teachers, students, administrators, and policy-makers.

Lessons from environmental education and related fields also provide potentially important support for mainstream international education and development efforts, such as UNESCO's Education for All initiative (which seeks to provide universal, equitable access to good quality education around the world) and the UN Millennium

---

[2] The nation's 22 indigenous groups receive little recognition within national political arenas, for example, and are often completely left out of national discussions of Costa Rican identity (see Stocker 2005; Mayorga et al. 2004; Minority Rights Group International 2008).

Development Goals (which include attention to both education and sustainable development). This is because much of what the existing research suggests is needed for 'good environmental education' is also central to 'good education' generally, including attention to student learning, effective teaching and learning strategies, and supportive learning environments (cf. Bangay and Blum 2010; Pigozzi 2007). The urgent need to address global environmental concerns such as climate change also underscores the need to better integrate the environmental dimension in all international educational development efforts.

Overall, this book has sought to both interrogate contemporary questions about the implementation of environmental education and also to address some of the gaps in existing research. I attempted to do this by examining the ways in which the production and shaping of environmental education is embedded not only within particular institutions, but also within daily lived experience and social interaction. In particular, I have used St. John and Perry's (1993) concept of an 'educational infrastructure' as a central metaphor for understanding the linkages between the physical infrastructure of schools and national policies, and wider networks of educational, social and cultural resources. I have also further extended this conceptualisation as a means to explore how knowledge is both disseminated and contested by community members. In this way, I have paired a broader understanding of 'educational infrastructure' with an anthropological interest in knowledge and power in order to critically assess the ways in which diverse individuals and groups interact with, negotiate and contest knowledge about the environment – as well as teach it to others. The focus of this research thus centred not on the composition of specific 'environmental messages' or a particular audience's acceptance of or resistance to them, but on the ways that diverse individuals actively negotiate and promote particular kinds of knowledge – for instance, through participation in educational processes in schools, social groups or organisations – and are both supported and constrained in their efforts by the social, economic and political contexts – local, national and international – in which they are situated.

Finally, given the complex global environmental challenges that face us in the contemporary world, there is a need for environmental education research to significantly expand its geographical focus. Work from Europe and North America continues to largely dominate the field, despite the fact that highly innovative practice and policy can be found around the world. Evidence and documentation (e.g. informal reports, project evaluations, policy documents) regarding these efforts can often be found at the country level, but may be difficult to access internationally. There is therefore a clear need for environmental education and related fields to broaden the horizons of collaboration and research to explore more diverse perspectives and contexts.

As an outsider to Monteverde myself, I cannot claim to directly represent local voices or perspectives on environmental education or sustainable community development. Indeed, the research suggests that these perspectives are in fact complex and highly diverse both locally and in the wider national context in Costa Rica. However, I hope that this book at least provides a starting point for further discussion and research about diverse experiences and understandings of environmental education around the world. In it I have attempted to give an accurate depiction of the active debates and discussions I witnessed while living in Monteverde, to provide a sense of the passion and dedication of the educators, conservationists, business owners and other residents who kindly allowed me to take part in their daily work and efforts on behalf of their community, and to give an account of the many examples of innovative thinking and practice that were shared with me, as well as of the key challenges the community continues to face.

References

Anderson, G., & Montero-Sieburth, M. (Eds.). (1998). *Educational qualitative research in Latin America: The struggle for a new paradigm*. London: Garland Publishing.

Bangay, C., & Blum, N. (2010). Education responses to climate change and quality: Two parts of the same agenda? *International Journal of Educational Development, 30*(4), 359–368.

Blum, A. N. (2006). *The social shaping of environmental education: Policy and practice in Monteverde, Costa Rica*. Unpublished doctoral thesis, University of Sussex, Brighton, UK.

Boyte, H., & Evans, S. (1992). *Free spaces: The sources of democratic change in America*. Chicago: University of Chicago Press.

DfES [Department for Education and Skills, UK]. (2006). *Sustainable schools for pupils, communities and the environment: Delivering UK sustainable development strategy*. London: DfES.

Dillon, J., & Teamey, K. (2002). Reconceptualizing environmental education: Taking account of reality. *Canadian Journal of Science, Mathematics, and Technology Education, 2*(4), 467–483.

Falk, J. (2005). Free-choice environmental learning: Framing the discussion. *Environmental Education Research, 11*(3), 265–280.

Fine, M., & Weis, L. (1998). *The unknown city: Lives of poor and working class young adults*. Boston: Beacon Press.

Fine, M., Weis, L., Centrie, C., & Roberts, R. (2000). Educating beyond the borders of schooling. *Anthropology & Education Quarterly, 31*(2), 131–151.

García Gómez, J. (2000). Modelo, realidad y posibilidades de transversalidad: El caso de Valencia. *Tópicos en Educación Ambiental, 2*(6), 53–62.

Gardner, K., & Lewis, D. (1996). *Anthropology, development and the post-modern challenge*. London: Pluto Press.

González Gaudiano, E. (1999). Otra lectura a la historia de la educación ambiental en América Latina. *Tópicos en Educación Ambiental, 1*(1), 9–26.

González Gaudiano, E. (2000). Los desafíos de la transversalidad en el currículum de la educación básica en México. *Tópicos en Educación Ambiental, 2*(6), 63–69.

González Gaudiano, E. (2007). Schooling and environment in Latin America in the third millennium. *Environmental Education Research, 13*(2), 155–169.

Grillo, R., & Stirrat, R. (Eds.). (1997). *Discourses of development: Anthropological perspectives*. Oxford: Berg.

Heimlich, J., & Ardoin, N. (2008). Understanding behavior to understand behavior change: A literature review. *Environmental Education Research, 14*(3), 215–237.

Huckle, J., & Sterling, S. (Eds.). (1996). *Education for sustainability*. London: Earthscan.

Jickling, B. (1992). Why I don't want my children to be educated for sustainable development. *The Journal of Environmental Education, 23*(4), 5–8.

Kollmuss, A., & Agyeman, J. (2002). Mind the gap: Why do people act environmentally and what are the barriers to pro-environmental behaviour? *Environmental Education Research, 8*(3), 239–260.

Læssøe, J., Schnack, K., Breiting, S., & Rolls, S. (Eds.). (2009). *Climate change and sustainable development: The response from education. A cross national report from the International Alliance of Leading Education Institutes.* http://dpu.dk/RPEHE and http://edusud.dk

Leitinger, I. (Ed.). (1997). *The Costa Rican women's movement: A reader.* Pittsburgh: University of Pittsburgh Press.

Lencastre, M. P. (2000). Transversalización curricular y sustentabilidad: contribución para la teoría y práctica de la formación de maestros. *Tópicos en Educación Ambiental, 2*(6), 7–18.

Levinson, B., Cade, S., Padawer, A., & Elvir, A. P. (Eds.). (2002). *Ethnography and education policy across the Americas.* London: Praeger.

Luzzi, D. (2000). La educación ambiental formal en la educación general básica en Argentina. *Tópicos en Educación Ambiental, 2*(6), 35–52.

Mayorga, G., Sánchez, J., & Palmer, P. (2004). Taking care of Sibü's gifts. In S. Palmer & I. Molina (Eds.), *The Costa Rica reader: History, culture and politics* (pp. 264–275). Durham: Duke University Press.

McKeown, R., & Hopkins, C. (2003). EE≠ESD: Defusing the worry. *Environmental Education Research, 9*(1), 117–128.

Minority Rights Group International. (2008). *World directory of minorities and indigenous peoples - Costa Rica: Overview.* http://www.unhcr.org/refworld/docid/4954ce31c.html

Ofsted. (2008). *Schools and sustainability: A climate for change?* London: Ofsted.

Palmer, J. (1998). *Environmental education in the 21st century: Theory, practice, progress and promise.* London: Routledge.

Palmer, S., & Rojas Chaves, G. (1998). Educating señorita: Teacher training, social mobility and the birth of Costa Rican feminism, 1885–1925. *Hispanic American Historical Review, 78*(1), 45–82.

Pellegrini Blanco, N. (2002). An educational strategy for the environment in the national park system of Venezuela. *Environmental Education Research, 8*(4), 463–473.

Pigozzi, M. J. (2007). Quality in education defines ESD. *Journal of Education for Sustainable Development, 1*(1), 27–35.

Reid, A., Jensen, B., Nikel, J., & Simovska, V. (Eds.). (2008). *Participation and learning: Perspectives on education and the environment, health and sustainability.* Dordrecht: Springer.

Reigota, M. (2000). La transversalidad en Brasil: Una banalización neoconservadora de una propuesta pedagógica radical. *Tópicos en Educación Ambiental, 2*(6), 19–26.

Rickinson, M. (2001). Learners and learning in environmental education: A critical review of the evidence. *Environmental Education Research, 7*(3), 207–320.

Rickinson, M., Lundholm, C., & Hopwood, N. (2009). *Environmental learning: Insights from research into the student experience.* Dordrecht: Springer.

Roth, E. (2000). Medio ambiente como transversal en la educación formal: Algunos apuntes en la experiencia Boliviana. *Tópicos en Educación Ambiental, 2*(6), 27–34.

Scott, W., & Gough, S. (2003). *Sustainable development and learning: Framing the issues.* London: Routledge.

Spindler, G. (1963). *Education and culture: Anthropological approaches.* New York: Holt, Rinehart and Winston.

St. John, M., & Perry, D. (1993). A framework for evaluation and research: Science, infrastructure and relationships. In S. Bicknell & F. Farmelo (Eds.), *Museum visitor studies in the 90s* (pp. 59–66). London: Science Museum.

Stables, A. (2001). Who drew the sky? Conflicting assumptions in environmental education. *Environmental Education Research, 33*(2), 245–256.

Stocker, K. (2005). *'I won't stay Indian, I'll keep studying': Race, place, and discrimination in a Costa Rican high school.* Boulder: University Press of Colorado.

Stromquist, N. P. (Ed.). (1992). *Women and education in Latin America: Knowledge, power, and change.* Boulder: Lynne Rienner Publishers.

Twombly, S. (1998). Women academic leaders in a Latin American university: Reconciling the paradoxes of professional lives. *Higher Education, 35*, 367–397.

Wals, A. (Ed.). (2007). *Social learning towards a sustainable world: Principles, perspectives and praxis*. Wageningen: Wageningen Academic Publishers.

Wax, M., Diamond, S., & Gearing, F. O. (1971). *Anthropological perspectives on education*. New York: Basic Books.

Appendix: Images of Monteverde

The road to central Santa Elena. Photo by N. Swetnam

Central Santa Elena. To the left are shops and a hotel, to the right is the Catholic Church and parochial hall. This is part of the only length of paved road in the district. Photo by N. Blum

158   Appendix: Images of Monteverde

View South from the village of Monteverde on the road to San Luis. Photo by N. Blum

The main building of the *Colegio* in Santa Elena. Beyond are a second school building, pasture for cattle, an experimental fish pond, and a building for chickens and pigs. To the right (just out of sight) is the office of the Santa Elena Reserve. Photo by N. Blum

Appendix: Images of Monteverde

Main entrance and administrative buildings of the Monteverde Reserve. Photo by N. Blum

Main offices and entrance of the Monteverde Institute, village of Monteverde. Photo by N. Blum

Index

A
Agenda 21, 2, 10, 21, 23, 145
Anthropology of development, 150
Anthropology of education, 13, 150

B
Belgrade Charter, The, 10

C
Cerro Plano, 56, 58, 60, 61, 126, 131, 133
Chamber of Tourism, Monteverde (*Cámara de Turismo*), 35, 87, 120, 121, 130
Climate change, 2, 7, 8, 146, 149, 152
Cloud forest, 41, 55, 59, 69–77, 84, 86, 94, 98, 128
Community development in Monteverde, 116, 122
Community of practice, 111
Conservation
    development in Costa Rica, 4
    early efforts in Monteverde, 83, 92
    'fortress' conservation, 87
    links to environmental education, 4, 10, 47, 138
    multiple-use strategies, 87
    scientific interest in, 10, 57, 89
Cooperativa Santa Elena, 90, 120, 124
Costa Rican Conservation Foundation (*Fundación Conservaciónista Costarricense*), 88, 119
Costa Rican Tourism Institute (*Instituto Costariccense de Turismo*), 54
Curriculum
    in Costa Rica, 148
    environmental education in, 13, 14, 38–40, 42, 43, 58–62, 66, 67, 69–78, 82, 91, 144, 148
    in schools in Monteverde, 58

D
Development, conceptualisations of, 116–124
Development education, 4, 10, 146

E
'Earth Summit' (UN Conference on Environment and Development), 10
Ecotourism
    in Costa Rica, 23, 35, 55, 65
    definition of, 35, 136
    in Monteverde, 7, 54, 85, 87, 128, 130, 134
Education for All (EFA), 54, 151
Education for sustainable development, 2, 4, 11, 145, 146, 149
Education policy
    historical development in Costa Rica, 27
    national system of assessment and examinations, 32, 42, 43, 59, 60, 64, 66, 71, 73, 74, 76
    structure of the Costa Rican education system, 26
Endangered species in Monteverde
    Golden toad (*Sapo Dorado*), 84
    Resplendent Quetzal (*Quetzal*), 88, 94
    Three-Wattled Bellbird (*Pájaro Campana*), 88, 119
Endemic species in Monteverde
    educational efforts related to, 93, 133, 142
    frogs, 68, 69, 134, 136, 137
    orchids, 129, 133, 137

Environmental education
   historical development in Costa Rica, 27, 90
   historical development in Monteverde, 90
   international history of, 2, 17, 45, 47, 66, 92, 143
   Latin American perspectives on, 2, 3, 148
   pedagogies, 139
   research on, 4, 11, 14, 15, 47, 104, 111, 147, 149, 152
   teacher education, 148
Environmental interpretation, 12, 98
Environmental policy
   historical development in Costa Rica, 24, 70, 147
   international, 24, 25, 66 (*see also* Rio Summit, Agenda 21, Rio Declaration, World Conservation Strategy)

F
Freire, Paulo, 2, 104

G
Gender
   in Costa Rica, 65, 102, 151
   in education, 4, 11, 151
Global education, 4

H
Ham, Sam, 12, 98

I
Indigenous groups in Costa Rica, 147, 151
International Environmental Education Programme, 10

L
Lave, Jean, 13, 110, 111

M
Millennium Development Goals (MDGs), 151–152
Ministry of Education, Costa Rica (*Ministerio de Educación Pública*), 29–32, 38, 39, 42–44, 46, 47, 59, 60, 62, 66–68, 76, 79, 97, 98, 149, 151, 152

Ministry of Environment and Energy, Costa Rica (*Ministerio de Ambiente y Energia*), vi, 34, 38, 40, 41, 44, 100, 135, 137
Monteverde Conservation League, 85, 86, 88, 90, 100, 105, 120, 126, 137
Monteverde Institute, 56, 85, 86, 88, 117, 122–126, 130, 131
Monteverde Reserve (*Reserva Biologica Bosque Nuboso de Monteverde*), 55, 61, 62, 70, 83–88, 90, 92–95, 97–99, 104, 105, 113, 114, 126, 129, 131, 133, 134
Monteverde, village of, 56, 58, 70, 85, 86, 88, 117, 118, 125, 126, 134

N
National System of Conservation Areas (*Sistema Nacional de Areas de Conservacion*), 36, 40
Non-governmental organisations (NGOs)
   in conservation, 26, 44, 57
   in Costa Rica, 25, 26, 41
   in education, 2, 25, 26, 41, 44
   in Monteverde, 26

O
Office of Environmental Education, Costa Rica (*Oficina de Educación Ambiental*), 38, 39, 42

Q
Quedar bien, 34, 56

R
Rio Declaration, 10
Rio Summit, 3

S
Santa Elena, 56, 58–60, 62, 64–66, 68, 69, 70, 76, 82, 83, 85, 86, 88, 90, 92–95, 100, 104, 105, 113–115, 120, 122–126, 130, 131, 133, 135, 137
Santa Elena Development Association (*Asociación de Desarrollo Integral de Santa Elena*), 113, 122, 124, 126

Santa Elena Reserve (*Reserva Bosque Nuboso Santa Elena*), 55, 66, 68, 76, 82, 83, 85, 86, 90, 92–94, 100, 104, 105
Schools in Monteverde
    Cloudforest School, 59, 69–77
    Colegio Técnico Profesional de Santa Elena, 58, 64–69, 85
    Escuela Cerro Plano, 60, 61
    Escuela Santa Elena, 60, 62, 95
    Monteverde Friends School, 59
Science education, 4, 44
Scientific researchers
    in Costa Rica, 17, 33, 36, 37, 47, 48, 64, 137, 143
    in Monteverde, 17, 57, 82–84, 87, 90, 92, 93, 105, 117, 128, 137, 142, 143
Social learning, 15, 78, 110, 111
Sustainable development, 2, 4, 6–11, 13, 15, 24, 26, 33, 37, 39, 40, 42, 44, 47, 48, 54, 60, 63, 71, 116–122, 124, 125, 130, 143, 145, 146, 148–152

T
Tbilisi Declaration, The, 101
Transversal themes (*temas transversales*)
    in Costa Rica, 42, 60, 64, 148
    in Latin America, 148
Tropical Science Center, 36, 84, 86, 87, 120, 137

U
UN Decade of Education for Sustainable Development, 145
Universidad de Costa Rica, 29, 31
Universidad Estatal a Distancia, Costa Rica, 31, 38
Universidad Nacional, Costa Rica, 31, 38

W
Wenger, Etienne, 13, 110, 111
World Conservation Strategy, 10

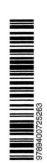